EMBRACE MAGNIFICENCE

A book filled with Natural knowledge From Cosmos to Cells

MOHAMED ABDULLA

INDIA • SINGAPORE • MALAYSIA

ISBN 979-8-89277-759-9

Mohamed Abdulla was born in Tamil Nadu, India. He pursued his bachelor's degree in mechanical engineering in 2013. He is a crash and safety Engineer who has worked for renowned automobile companies like General Motors, FORD through TCS, and Rle India Private Limited. Since childhood, he has been passionate about writing, often filling his hands with ink. He has written many articles and poetry in his mother tongue, Tamil. His notable works, 'Minnalazhagae' (Lightning Beauty) and 'Story of a Sculpture,' were well-received and highly appreciated by his school and college tutors, earning him respect and rewards. Mohamed is inclined towards spiritual and scientific subjects and loves teaching science to students. He firmly believes in the famous quote of the legendary scientist Albert Einstein, "Science without religion is lame; religion without science is blind."

He was born into a family of five siblings, being the eldest. Despite a meagre financial background, his parents, Habeeb Rahman and Alima Beevi, nurtured their children with love, knowledge, and peace. His parents, along with his better half Sulaika Jasmine and his entire family, provided great support during times of hardship and joyful milestones. His friends and colleagues have positively influenced his life, especially his childhood friend Noorulhasan, MBBS MD Psychiatry (NIMHANS), who is a significant source of inspiration.

Mohamed's aspiration to launch a book was fulfilled with the support of his friend, mentor, and brother Shivathmaj M. Shenoy, a college student and author of 'Journey Made Awesome' and '5 Great Indian Kings and the Legendary King Maker.' You can reach the author of this book at the email address below:

pennameabdulla@gmail.com

Contents

Prologue

I am dedicating this book to all the humans who strive to become more than human.

You were made for this; Go forth and Conquer.

Leave not your life without leaving a legacy.

May Almighty bless us all

– Mohamed Abdulla

Foreword

Do you know? The rainbow you see is actually a full circle, the bacteria present in your mouth more than the number of people in our planet, our brain has sufficient electricity to power a light bulb like much more fascinating details and science concepts, I am presenting this book *'Embrace Magnificence.'*

The prime motive of this book is to enrich the natural knowledge surrounds us. This book will gift you a special lens so that you can see this world, a few layers deeper. There were times, people travelled miles departing from their families, sacrifice their lives to acquire knowledge. They have inherited the same to their progeny and successive generations.

Nowadays Knowledge is within reach of every hand. In this book, I have elucidated familiar concepts in a non-familiar way with fascinating details. Every awestruck moment while you are reading this book, it is not because of this book. But because of the way things have been created and functioning in our universe. *'We have only one life to appreciate the grandness of the universe'* is one of my favourite quotes said by the phenomenal man, Stephen Hawking. *'Knowledge of nature and creations unleash the optimist inside you.'* Different concepts discussed in this book, takes you to different worlds. It kindles your curiosity, makes you feel special, unique, and more importantly it keeps the student in you alive. Universe has no bounds and so the knowledge. Whatever discussed in this book is just a drop of an ocean. Nowadays lack of knowledge is not a problem, but lot of knowledge is the problem.

Knowledge without action is useless. There are lessons we can imbibe from nature which I have touched in this book. Even if we start implementing a single activity in our daily lives with utmost dedication, you will find how beautiful the doors are opening in your life from the direction which you have least expect it. May Almighty make our life, be an example rather than a warning to the people. Have an incredible journey of enriching natural knowledge, hope this book will serve as an elixir of your soul.

Preface

Surely in the creation of the heavens and the earth, and in the alternation of night and day, there are signs for people of understanding.

They are those who remember Allah while standing, sitting, and lying on their sides, and reflect on the creation of the heavens and the earth and pray, "Our Lord! You have not created all of this without purpose. Glory be to You! Protect us from the torment of the Fire."

Holy Quran (Chapter 3: verses 190-191)

ONE

Embrace Magnificence

2020, a hard hit to the human race. It unleashed a dreadful disease and shook the entire Earth. People were homebound, directionless, fearful of facing the terrible virus and shackled from their fundamental freedom of going out and engaging with the environment. This year left a mark in human history in its own way.

Amidst all this chaos, I was thinking of digging about the people in the far worst phase of this worst situation. So, it may bring solace to my soul and comfort to my heart, and it may help me find the blessings that I failed to spot within myself. It happened to me to meet my friend, who was the victim of the virus. He confidently combated with it and successfully overcame the worst situation. Now, he is back to his fullest health. The experiences he shared with me enlightened me a lot. It channelled me to scribe this chapter, *"Embrace Magnificence."*

He started narrating his bitter experience of those quarantine days. He said, "For two months, I was in a room filled with medical equipment, a few kinds of stuff for my basic needs. I was given a few kinds of food and drinks that were necessary for me to survive. There was no communication with the outside world. I was ignorant of all outside affairs. Agony gripped my heart all the time. With sheer confidence and continuous medical treatment, I have won the battle. What's more exciting is the day I came out of the room post-discharge. It was like I was reborn again. I felt like I was seeing the world for the first time with full consciousness."

"The warmth of the Sun rays striking my face, dazzling white clouds upon my head, the sight of twinkling stars during the nighttime, and green trees surrounding us. Indeed, it is a world of fascination." He stopped with this exciting experience.

I Sensed a source of enlightenment in his speech. I cast my thoughts on his sources of excitement. I have tried to plant consciousness in all my ordinary things. The broadness of the world, the deep blue colour of the sky, the sounds of rivers, lakes, and seas, the rain which brings life to Earth, flowers of various kinds, Jasmine, roses with their pleasant aromas, butterflies with their fascinating colours. When you start feeling the magnificence surrounding you, your heart involuntarily thinks of the creator who created this wonderful phenomenon and put it together to work interactively. We get awestruck by the man-made steel bridges, but we are oblivious to our human bone, which has the strength of steel.

It is very important to understand the existence of all these things. We should not look at these things as an act of habit but with full consciousness. When you start experiencing and understanding things around you, your astonishment will grow even more. It will topple your life from misery to joy. It will make your ordinary day into an extraordinary one. Let the energy around you flow through you.

I have done my best to present a basic understanding of these marvellous creations with their fascinating details. I strongly believe every piece of information you learn enables you to embrace this magnificence even more.

COSMIC GIFT

Everything you see—the book you are holding, the armchair you are sitting in, the house you live in, your mother, your father, in short, every object you see, including objects you cannot see like cells, microorganisms, atoms, molecules, and the giant objects which you cannot imagine the size of, such as stars, planets, and galaxies—all belong to the universe.

The universe is everything; the origin of the universe is the origin of everything. Have you ever thought of how it all began? How did the universe come into being? This question holds a mind-blowing answer of infinite power, wisdom, and energy.

The story of the origin of the universe starts at a time when time does not exist. About 13.7 billion years ago, a tiny hot particle with infinite density exploded. This explosion is called the Big Bang. Over the last century, with a series of scientific experiments, observations, and scientific calculations using advanced scientific technologies, scientists have proved beyond any doubt that the universe has a beginning.

Our universe is continuously expanding and extends in all directions like a balloon inflating. If we could travel back in time, the universe must have exploded from a single point. This huge explosion happened in unimaginable ways. You cannot imagine this 'Big Bang' as an ordinary explosion with a giant ball of fire and smoke around the entire space because it happened at a time when space and light did not exist as a result of this explosion, matter, time, space, and energy came into existence. Space continues to get bigger and cooler. At no time does its size become big; for example, the average distance between two stars is 2 trillion miles.

We have witnessed thousands of explosions on the Earth's surface, none of which brought constructive results. All the explosions we have witnessed brought destruction, disaster, and disorderliness to the environment. However, the Big Bang explosion was not an ordinary phenomenon. It brings constructive results, perfect order, and harmony to creations. Because of this, life came into existence. It is a well-planned, meticulously devised, flawless creation from the creator of the universe. Let's say you are imagining a huge explosion happening by burying a dynamite explosive underground.

As a result of this explosion, a huge, beautiful palace emerges from the Earth with eye-catching paintings, colourful windows, and marvellous design, all with perfect dimensions. Could you believe this scenario? The answer is 'certainly not.' Likewise, this Big Bang explosion, which brought stars, galaxies, the solar system, and Earth as a planet of life, did not occur as a matter of coincidence. There is infinite wisdom, mighty knowledge, and a perfect plan behind it.

The expansion rate of the universe is not random. There lies a delicate balance in the rate of expansion in the universe, maintaining a perfect state of equilibrium. If even a one-in-a-billion times difference occurs in this equilibrium, the universe would not come into existence.

Let's say there is a slowness in the expansion rate of the universe; due to gravitational effects, everything in the universe would start colliding violently. On the other hand, if the universe expands a little too quickly, all the cosmic matter in the universe will escape into space and vanish away. Countless conditions must be satisfied to maintain this equilibrium.

Everything in space depends on flawless laws and extraordinarily delicate balances set into motion by the creator of the universe. This is how the universe began. This is the story of the Big Bang.

There is something common between all of us. We all love watching the sky during nighttime hours. I still do it often with my kid; we go to the terrace at night, spread out a mattress, and lie on it for some time. Staring at the stars, Moon, clouds, and the rare sight of comets occurring once in a blue Moon brings delight to our hearts.

The vastness of space surpasses our human imagination. Scientists have proved that no matter how fast and how far you travel, there is no end to the universe.

The Sun, one of the medium-sized stars in the universe, has a diameter 103 times that of the Earth. Yes, you can fit more than one hundred Earths inside the Sun. The larger a star, the faster it burns.

If the Sun, which provides life to Earth, were ten times larger than it actually is, it would have died 10 million years after its formation, and we would not be here right now.

The distance of the Sun from our planet is very accurate. If our planet's orbit were slightly farther from the Sun than its actual distance, everything on Earth would be covered with ice. If the Earth's orbit were closer, then everything on Earth would evaporate. The Sun is located in the universe, and it is ideal in size and distance to support life on Earth. Countless systems operate in the universe while we remain unaware of them. Everything is under the control of boundless intelligence.

The Sun is the largest body in our solar system, yet it is one of the medium-sized stars in the vast universe. There are more than 250 billion stars in the Milky Way galaxy, just like our Sun. The Milky Way is the galaxy where we live. There are many more massive stars than the Sun. No two stars are the same. Each star is unique, with its specific temperature, size, colour, and brightness. An interesting fact is that even the Milky Way galaxy is relatively small. There are more than 300 billion galaxies present in our space. It goes beyond our human comprehension. The vastness of space is always restless.

Scientists who have uncovered the size, distances, and balance of the universe have asserted many times that it indicates an obvious design. But science often serves as a disservice because, with knowledge, we believe we have control over things. There is a universal power that creates and sets everything into motion with a delicate balance, having control over everything. Despite the world being so vast, like you are a minuscule grain of sand compared to the immense scale of the world, God still loves you. He is in all of our lives now and forever. Do you think the one who set the space with impeccable design and delicate balances does not add meaning or purpose to our lives? Acknowledge him and stand firm in your faith.

There is a purpose, a meaning for your existence; you are not taking a single breath in vain. Compared to the clock of the universe, humanity's time span was probably just a blink of the cosmic eye. We have infinite potential; we can accomplish so many things, yet we are a speck of dust whose existence is nowhere compared to the grandness of our universe.

We can achieve whatever we are ambitious about, but whatever heights we reach, we should learn to acquire the virtue of humbleness. Let us humble ourselves under the mighty hand of the Almighty. He himself lifts us all up. He is in our lives all the time. It is never-ending grace.

So, never stop staring at the stars. It reminds us of how we are insignificant and significant at the same time. It reinforces our belief. Do not forget that there is always a ray of hope.

A GLIMPSE OF OUR HOME

I remember a time when my brother entrusted me with a huge responsibility, which I was not accustomed to. My brother is a fish enthusiast; he enjoys keeping fish. He has a small aquarium at our home, which he used to care for with utmost joy. One-time, he had to go on a long trip, so he handed over the responsibility of caring for the fish to me, and I gratefully accepted. I began feeding the fish every day. After a week, I decided to change the water, so I dismantled the aquarium setup and placed the fish in a small bowl of water. It was the first time I was setting things up on my own. This setting process provoked deep thoughts about life in my mind.

The aquarium setup provides all the appropriate conditions for a fish to survive: a protective cover, a motor for ventilation, a thermostat maintaining the proper water temperature, a filter system continuously cleaning the water, and a continuous food supply. All these settings support the life of the fish in the aquarium. However, the fish within the aquarium are unaware of this artificial environment; they believe they are living in a natural environment.

They are unaware of the people who do all the vital setup, and they do not know who supplies the food that suddenly appears on the surface. Yet, it is obvious that the owners of the aquarium provide everything the fish need. Similarly, our life on Earth requires more delicate systems than life in an aquarium. We are people of intellect; we should not spend our lives in ignorance like fish in an aquarium. It is our primary responsibility to understand the creator through His creations; let's examine the balance, impeccable design, and harmony on Earth that make it a liveable place for us. In this way, we can witness our creator's infinite might.

Just think for a moment about what is necessary for a human to survive. All the necessary conditions and details for living, both known and unknown at this moment, are naturally available on Earth. These necessary details are interrelated. Among all the planets in our solar system, Earth is the only planet suitable for life. The light coming to us from the Sun, the water we drink, and the food we enjoy are extremely suitable for our lives.

Let's consider the structure of our Earth. Earth was created about 4.5 billion years ago from a mass of rocky debris rich in iron orbiting the Sun. Imagine a mango fruit; picture it as spherical in shape, similar to our Earth.

- The centre of the Earth is made of iron. This metallic core resembles the hard stone at the centre of a mango.
- Next is the hot, mobile rock of the mantle, which is like the juicy flesh of a mango.
- Then, at the top is the Earth's rocky crust, which is like the skin of a mango. This top layer of Earth is called the crust, and it is the reason behind the formation of our magnificent mountains.

The Earth's crust, the surface on which we walk and build our houses safely daily, actually moves on a layer called the mantle, which is denser than the crust. If there were not a system in place to keep this motion under control, continual shocks and quakes would occur on Earth, making it truly uninhabitable. These gigantic mountains have come about as a result of movements and clashes of huge plates that make up the Earth's crust: when two of these plates collide, one usually slides under the other.

The plate on top is pushed up, forming mountains, while the plate at the bottom proceeds under the ground and forms a deep protrusion. This means that mountains have a deep downward protrusion as large as those visible on the surface. In other words, mountains are firmly rooted in the Earth's mantle, preventing the crust of the Earth from sliding. Simply put, just compare mountains to nails firmly holding pieces of wood together.

We all know that temperature reduces with an increase in altitude; due to this, very high mountains are usually covered with snow. There is a delicate equilibrium present in the thickness of the Earth's crust. If the thickness of the Earth's crust is higher, more oxygen from the atmosphere is transferred to the Earth's crust. If the thickness of the Earth's crust is less than ideal, then volcanic activities and earthquakes will occur more often.

Let's take a look at another important aspect of the Earth that makes our planet liveable. When you go outside, the Sun's rays strike your face without bothering you, conveying pleasant warmth and heat for your benefit. But in fact, our Sun is like a deep pit consisting of red gas clouds, made up of a whirlpool of giant flames that gush from the surface to millions of kilometres away—giant tornados rising to the surface from the bottom. It could be fatal to mankind, yet our Earth's magnetic field and atmosphere filter out all the harmful rays from the Sun before they reach our Earth. This serves as a protective shield for the Earth. Earth's atmosphere is made up of seven layers, each composed of different gases in perfect harmony with each other. The Earth's atmosphere maintains the correct proportion of gases. For example, for everybody to breathe continuously until we die, it is necessary that the balance of gases in the atmosphere is just right. Even slight changes could be fatal to mankind.

Earth's atmosphere is made of 77% nitrogen, 21% oxygen, 1% carbon dioxide, and other gases. Oxygen is a very important gas to live. This oxygen ratio is maintained with an extremely delicate balance. We, human beings and animals inhale oxygen and exhale carbon dioxide. Plants do quite the opposite of this. What if we humans and animals carried out the same process as plants? In that scenario, the oxygen level in the atmosphere would rise. If the atmosphere is filled with oxygen, even a small spark would cause a huge fire. Consider someone working in the kitchen left the cylinder open. What would happen? The entire room would be filled with gases, and even a small spark would cause a big explosion. On the other hand, if plants did the same process as we do, we would run out of oxygen, and gradually, we would cease in suffocation. There is an extremely delicate balance that maintains the oxygen ratio. In order to survive life on this planet, countless equilibria need to be maintained.

Think about when we stand and begin walking; we feel no pressure in an upward or downward direction. Each time we engage in our daily activities, we are completely unaware that we are resisting a powerful gravitational force. The most important reason for this is the size of our Earth. If it were just slightly smaller, gravity would be far weaker, the planet's atmosphere would fragment and disappear, and we would be unable to remain stable in the world. If the Earth were larger, gravity would increase considerably, and various poisonous gases would make our Earth lethal. Even if we managed to protect ourselves from these gases, we would be unable to move. But such a problem never arises because the Earth's size has been determined in a manner that makes human life possible. Thanks to the infinite power and wisdom, our master has arranged matters so that human beings have no need to eat or drink gases in specific proportions.

Every night, I love to watch the Moon. It is very beautiful and always a source of delight. A few hundred million years ago, an Earth-shattering event occurred. As a result of this, a satellite body moving around our planet called the Moon. It is the natural satellite of our Earth, the largest source of light at night. The Earth and the Moon are always moving away from each other. If the momentum and the gravity were to go out of balance somehow, the Moon would not stay in orbit; it could keep going away or fall to the Earth. Look how meticulously the balance has been established here.

Let's talk about another profound balance nature exhibits. We discussed the atmosphere and how the gases present are proportionate to maintain life on Earth. The atmosphere is made of light air. This doesn't mean the atmosphere has no weight. Actually, these layers of air, kilometres thick rising above us, are very heavy. This atmosphere exerts tonnes of pressure on every one of us. This is what we call air pressure. Now, you may wonder why we aren't crushed. We aren't crushed because our bodies are created with the strength to withstand the weight of the atmosphere. If the pressure were low, we would no longer exist in this environment. Without this pressure, the blood circulating rapidly in our body would exert tremendous pressure on our veins; unless balanced by atmospheric pressure, veins would burst due to high-pressure. Therefore, it's not possible for human beings to live on a planet like Mercury, lacking an atmosphere. There is an atmosphere on our neighbouring planet Venus, but the pressure on Venus is 99 times greater than Earth's atmospheric pressure, which does not provide suitable conditions for human life.

From all the above details, we can understand how well-balanced our universe and planet are. Nature has a mechanism of striking balance, oscillating between extremes. Moderation is the key because it aligns with nature. Remember, whether it's rise or fall, gain or loss, pleasure or pain, victory or defeat, always maintain balance. Do not halt, do not stagnate; always keep moving. We can reap all the benefits only if we stay between two extremes. Work hard but not overly hard. Remind yourself to work in order to live and not live in order to work. Your job is only a part of your self-worth, and you have other achievements in life. Save money when you have a job, but once in a while, take a break and enjoy the simple joys of living. Plan for the future, but don't plan too far ahead. It's important to balance and diversify all aspects of life, not to spend an excess of time or energy on only one aspect.

Let your priorities strike a balanced life; more importantly, be present in your life. Live consciously moment after moment. If you are in a rose garden, don't forget to smell the roses. Personally, I need to moderate my time management. I feel most of my time is dedicated to working. Now it's your turn to find out your own. Think about where you can moderate yourself. Every individual should avoid extremes and focus on having a life committed to balance and wholeness. Find your own middle ground between excess and deficiency in all things; that's where your best benefits hide.

TASTE OF SKY

Nowadays, the act of phubbing affects social relationships. I believe everyone is aware of the word 'Phubbing' these days. For those who are not aware, 'Phub' is a word recently coined by a dictionary organisation. It's a composite of two words (Phone + Snub) - snubbing or avoiding someone you are talking to in order to look at a cell phone. This phenomenon seems harmless, but research is finding that it significantly affects personal relationships.

Think back to our childhood days before the advent of cell phones; we felt more connected with relationships, nature, and the natural world rather than the virtual. I still remember when I was a kid, I would look up at the skies and watch the clouds go by. This was common for every one of us in our childhood days. But in today's world, even young school-going kids are found busily chatting away on their phones. The addiction to mobile phones is affecting their attention. It is our responsibility to teach them to detach from the virtual world and spend a little of their precious time with nature. Just pass on your fondness for gazing at clouds to your children; it teaches them a lot. Clouds are the imaginary playground for a kid's imagination; it enhances their creativity.

Clouds are made of round water droplets or ice crystals floating in the sky. The Sun's heat causes water on the Earth's surface to evaporate. This water vapour is carried by air to high altitudes where it encounters cool air and turns into tiny droplets, forming clouds. An average rain cloud holds about 300 million kilograms of water. The tiny droplets in the clouds join and turn into larger drops; when that happens, gravity causes them to fall to Earth - that is what we call rain. NASA also studies clouds on other planets. Mars has clouds that are similar to those on Earth, but other planets have clouds made of substances like ammonia instead of water.

Water evaporating from the oceans contains a high ratio of salt. When it falls on land in the form of rain, it is pure and clean. If rainwater were very salty, it would wither the soil and plants, resulting in the death of all living things. Our life on Earth would soon come to an end. Certainly, this is the greatest blessing from our creator.

Rain falls to Earth in a certain measure. Each year, an average of 505 trillion tonnes of waterfalls on Earth. This amount does not alter from one year to another. The evaporation and condensation are in equal amounts - this is what we call the 'water cycle.' Maintaining this measure artificially would be impossible, even with all the advanced technologies available. A minor change in the water cycle could cause significant instabilities in nature in a very short time.

Rain brings everything in its own beautiful way. A rainbow is one of the countless beauties of nature that we all might have seen at least once in our lives. The enchanting colours and shape are quite astonishing. Just a glimpse of a rainbow makes all our miseries, worries, and problems seem to dissipate at that moment. You fall in love with nature. A rainbow is an arc of seven colours lined up one after another.

It's actually a trick of the light. White rays of the Sun are multi-coloured; once sunlight passes through a raindrop, the raindrop acts as a prism, separating the white rays into seven colours which make up white light. This separation involves a process called Total Internal Reflection.

Sunlight undergoes reflection and refraction inside the raindrops at different angles, refracting out the different colours within white light at different angles, creating the spectrum. The basic colours of the spectrum are red, orange, yellow, green, blue, indigo, and violet. One interesting fact about a rainbow is that we always see them as semicircular. However, the rainbow is actually a circle. It's not possible to observe all of these colours from the ground, which is why we see the rainbow as semicircular. Only from an airplane can we see them as a full circle. The centre of the rainbow circle is always a point exactly opposite the Sun. As the Sun rises higher, the rainbow also moves higher to remain at the same level as the Sun. These are the creations of our Lord so that we can enjoy the pleasures and beauties of our Earth.

The rain from the sky brings life to our Earth; it brings forth plants of different kinds. Plants are essential for the survival of human beings. Just think for a moment, what is the basic need for human survival on this Earth? Water, oxygen, and nourishment. The green plants around us ensure all of these basic needs. There are a lot of other balances on Earth maintained in equilibrium by these plants. As we all know, the main source of energy for life on Earth is the Sun. However, human beings and animals are unable to make direct use of solar energy because their bodies lack the system to use this energy as it is.

For this reason, solar energy can reach human beings and animals as usable energy only through food produced by plants. When you sip tea, you are actually sipping solar energy. When you eat food, you are actually consuming solar energy. The strength in our muscles is really nothing other than solar energy in different forms. Plants store this energy for us in molecules in their bodies by carrying out a complicated process. Due to their unique cell structure, plants use solar energy and convert it into energy that people and animals can absorb through nutrition. This collective process is called photosynthesis. Every structure in plants has been specially planned and designed.

But sadly, we are not realising it; we cut down trees and pollute our environment without any guilty conscience. We have to learn to economise things; we should not waste resources simply because we can afford the bill. We may not need it, but other people who share the same globe with us do. It is a sign of ingratitude to waste resources. Imagine I need a tree to make furniture; I should not go and deplete the entire forest for the sake of building furniture. Every one of us should ask ourselves how many trees we have planted for future generations. We should use resources only when they are necessary and preserve the rest for our future. We are surrounded by the greatest blessings of our creator. If we want to live a life of gratefulness and show gratitude to our master, we should stop wasting our natural resources.

We must realise that God appointed us as caretakers of our Earth. Despite the damage we cause, our Earth is resilient; it is still rich in resources. This resilience is an excellent virtue we must learn from our life planet. Resilience is not just a fancy word; it symbolises battle. We all get humbled at some point in time; we are all bound to face situations that go from bad to worse. Take your first step and start moving again and again. Resilience is the ability to get up again when we face a fall. The difference between ordinary people and those who turn their lives extraordinary is the way they deal with failures and setbacks in life. At every fall, they refuse to remain down; they refuse to give up. Failure is just feedback to them but never final. We must learn to stay as resilient as our planet Earth. Despite deep scars and countless blows, it replenishes life again and again. Stay resilient. Both we and our planet should thrive together.

TWO

Worlds Luxurious Things Are Free of Cost

One Saturday evening, I was in a shopping mall along with my wife and my 3-year-old child. It was a kids' section filled with toys. I was asking my child to see what he liked; he pointed to a particular toy. I was about to purchase it when suddenly he stopped me from grabbing the toy, dragged me outside, and asked for the electric car placed outside. I know it's worth more than 10,000 rupees, which is beyond my wallet's capacity. I calmly told him, "We will buy it next time." But he was stubborn about his choice, and then I told him quietly in his ear, "It costs a lot, Dad doesn't have that much; we will buy it once I have enough money." He understood and told me that I had to buy it when I had enough money. Then, we left the mall with another little toy.

The next day, I was playing with my child, as we often do with a pillow fight. We assigned ourselves superhero characters; he loves to play Spider-Man, and I was playing Hulk. As we were playing, suddenly, he stopped and asked me to wait. He went to a nearby corner of the room and said, "Sir, my spider web has finished; please give me some more web for five rupees," as if he were speaking with a shopkeeper. Five rupees is the maximum amount of money for him. I was amazed by his creativity, imagining a scenario where Spider-Man runs out of the web while chasing an enemy and has to renew his web power with money.

Another day, when I had a difficult conversation with my wife, he sensed his mum was unhappy. Slowly, he went over to her and whispered, "Mom, don't worry, I will bring Dad's wallet without him noticing. You can buy whatever you want." We all laughed when we heard what he said. It lightened the situation. Such moments happen to all of us. The notable point here is that from childhood onwards, we are accustomed to relating the value of money to everything. It's hard-wired within our brains. We think all desires and necessities in life are fulfilled by money alone. With money, you can go on vacation, but happiness? With money, you can buy a bed, but sleep? I am not saying money is nothing. Money is important, but there are many greater things beyond money present in our lives. We have to learn to experience and enjoy everything.

We all know Steve Jobs, a billionaire who passed away at the age of 56. His last words, while lying on his sickbed, were, "I took little joy in my life, but in other's eyes, my life is an epitome of success. I am recalling my life at this point, all the recognition and wealth which I took so much pride in, have paled in the face of death. Material things lost can be found. But one thing can never be found if lost, and that is LIFE. Treasure the love for your family, spouse, and friends. Treat yourself well. Cherish others. As we grow older and wiser, we realise that wearing a £300 or £30 watch both tells the same time; whether we drive a £150,000 or £30,000 car, the road and distance are the same, and we reach the same destination; whether the house we live in is 300 sqft or 3000 sqft, the loneliness is the same.

Material things don't bring you happiness. Happiness is within you. Don't educate your children to be rich, educate them to be happy. So when they grow up, they will know the value of things, not the price. Enjoy a healthy life God loves you. It's up to us to gift ourselves, the joy of nature's luxury by planting consciousness. Moment by moment enjoys your luxuries at no cost."

GOOD ACCIDENTS

Accidents: how does this word sounds to you, do you feel something negative, do you think you have read something unpleasant. But I don't think so, because in my day-to-day life it is very common. Accidents are getting entangled with my routine job. Yes, I do accidents every day, without crashing cars I would not be paid. Are you getting puzzled about my job, don't worry I'll take you a short trip of my profession it will make you understand what I'm doing and with my job what I am trying to explain under this topic. I am a crash and safety Engineer.

Working for automobile companies, around the globe for the years. My job is to check the crashworthiness of a vehicle using multiple advanced simulations. So what is the crashworthiness of a vehicle or a car? Crashworthiness of a vehicle is the ability of a vehicle structure to save its passenger when it encounters a crash. For example, if an accident happens the vehicle must absorb all the energy, it must undergo deformation so that the passenger inside the car must be saved. You might have come across the safety rating of many cars like 4 out 5, 3 out of 5, it depends upon the crashworthiness of the vehicle. There are many factors involved in the cessation of the life of a passenger. It is not necessarily physical damage like head injury, chest deformation is the only cause for the person to die during an accident. Broadly speaking there are three ways an individual can succumb to death during a car accident.

1. By means of physical damage like head injury, chest impact, ankle breakage, etc.
2. Damage is caused within the internal organs of a person, like a brain getting bombarded with a skull, or heart getting an impact with a rib cage, which causes sudden death due to violent impact within a short duration of time.
3. Passing of heavy energy inside the human body due to impact, which the human body could not absorb. As a result, the passenger succumbed to death.

There are many safety factors employed in vehicles to reduce damage caused by the first two scenarios, such as airbags and seatbelts. However, when it comes to managing energy, the vehicle's structure, materials, design, and position all work interactively to protect passengers from the transfer of energy to the human body. What is this, and how does it cause death? For example, when driving a vehicle at 100 km/hr, you are also moving at that speed along with your vehicle. During an impact, due to inertia, we cannot stop ourselves from colliding with the car's dashboard. Airbags are employed to decelerate our movement and protect us from head impact with the dashboard. Seatbelts are designed to mitigate chest deformation by retracting the passenger from sudden movement.

A crucial aspect of science is understanding energy's existence. When a car is moving at around 100 km/hr, it possesses an enormous amount of kinetic energy that does not dissipate easily. During an impact, this kinetic energy is converted into internal energy within the car. This conversion leads to severe deformation of the vehicle as it absorbs the energy internally. It is critical to prevent this energy from passing through the human body, as the human body cannot absorb the sudden surge of energy and the severe vibrations caused by the accident. For instance, even if the energy passes through the human body in just 3 milliseconds, it is enough time for a person to die. To clarify, the time mentioned is not in minutes or seconds but in milliseconds. One second equals 1000 milliseconds.

Consider how little time we have to save the passenger in such situations. That is why we design the front end of cars to absorb all the energy by undergoing deformation before it reaches the passenger compartment. Most fatalities in accidents occur due to this energy phenomenon rather than solely physical injuries. While airbags and seat belts can protect against physical injury, the vehicle's design and materials must cooperate to absorb internal energy effectively.

A strong car is not necessarily a safe car; if it is too strong, it may transfer all the energy to the passenger, resulting in fatalities. You may have seen people protect their vehicle's bumper by installing a front bumper guard. We often believe this will shield the vehicle during a crash, but in reality, the bumper absorbs 50% of the energy. Preventing bumper deformation can lead to fatal outcomes for passengers. During an accident, multiple mechanisms must work interactively without delay to save lives.

Did you know that the complete duration of a car accident does not typically exceed 60 to 70 milliseconds?

Everything happens within this unimaginably short span of time. To save passengers from collapse, we must work more precisely, without any lack of time. In many real-life scenarios, doctors often say, "If you had brought this patient 5 minutes earlier, we could have tried our best to save him." But nowadays, we are far ahead in utilising time from seconds to milliseconds. Seconds are no longer sufficient for us, even to mention. Our world is evolving every moment. We are going deeper and deeper into utilising time. We have to do this because there is no way you can buy back your lost time, even if you spend your entire wealth. Even a millisecond delay would end in disaster.

In an accident scenario, when an impact happens, sensors in different regions of the vehicle should immediately recognise the severity of the impact. If it is severe, they should send a signal to the airbags, seatbelts, and all the active safety systems in the vehicle. The safety features should start deploying in no time to save the passenger from injury and death. The vehicle body should absorb all the energy during the impact and neutralise the energy before it hits the passenger compartment. So, the passenger will be safe. Everything happens within 20 to 30 milliseconds.

If there is even a single millisecond delay, it is hard to save the human.

It has happened many times; due to technical issues, airbags won't deploy at the right time, and many people face death. Even the world's best airbag manufacturers have recalled their products several times due to malfunctioning airbags.

What I found profound in this process is the importance of time. How time plays a major role here. If we expand the time duration of the travelling from the front end of the car to the passenger compartment a little, it's a high possibility of saving the passenger. Longer the duration of time, lessen the impact faced by the human. I will explain this with a simple example. Consider person A driving a car with a speed of 56 km/hr, and person B driving another car with the same speed.

On the way, both encounter a barrier. Person A applies panic braking at the instant and stops the car. Person B applies the brake gradually and stops the car. Now both are saved from hitting the barrier. But person A received more impact than person B while applying the brake. Because due to panic braking, person A didn't give much time to stop the car, so the impact he gets is larger than that of person B, who gives time to stop the motion of the vehicle. So if we expand the time duration, the impact will be less.

Regarding the expansion of time, I am not talking about a few minutes or a few seconds. What we need is just 20 to 30 milliseconds of time if we could manage the travelling of energy during the impact before reaching the passenger compartment. There is a high possibility of saving humans during an accident. Within the shortest span, all the safety features like airbags, seatbelt, and front vehicle structure must coordinate interactively to save the passenger. Airbags and seatbelt are used to save the passenger from physical injury.

In order to save the passenger from energy waves from the impact, we designed the vehicle in a manner that it should crumble and absorb the energy from the impact and save the human. So, if you see any car of the front end crushed enormously due to an accident, do not think the human inside the car might undergo severe injury. Don't think like that because it is vice versa. The car front structure absorbs the energy a lot by undergoing severe deformation, so there is a possibility for the human to survive unless he wears a seatbelt and airbags deployment happened properly.

The primary function of all these safety features and design of the vehicle structure is to delay the time of the impact energy to get into the human body. You cannot assign a cost to human life and so time. Ultimately, you can save lives by saving time. Every breath we are our time span is getting shorter. It is up to us to use it or waste it. Gone time never be earned again. One humble request to my readers, behind the safety features of the vehicle, thousands of brainpowers, tonnes of brainstorming, lives of many intellectuals present to save the human.

But with your little ignorance or lethargic manner, don't spoil your precious life. Don't forget to fasten your seatbelts while before starting your car. We are scrutinising all the possibilities to expand a few milliseconds of time. Don't take your time for granted, which is the greatest blessings from our creator. Sculpt your life by sculpting your seconds. Time is free, but it is priceless.

There are uncountable, unimaginable numbers of events taking place every second of our life. While you are reading this book, 100 billion processes taking place in your eyes every second. 8 million cells in your body died, 8 million new cells are getting produced to replace them. There are more than 2.5 million red blood cells are manufactured in our body. Our heart is pumping three cubic centimetres of blood. There are thousand nerve cells are produced in a fetus in its mother's womb, there are innumerable phenomena taking place inside and outside of you, How many of us aware of it. Each phenomenon is an independent miracle. Most of us are unaware of the fact that our lives depend on thousands of different processes taking place in a particular sequence and amazing order.

Each one of the links in this marvellous system works in perfect harmony and interdependence with all the other processes. The absence of a single link brings destruction in unthinkable ways, eventually leading to the destruction of our planet Earth. Our world survives due to the constant maintenance of both known and unknown delicate balances, which include the rate of expansion of the universe, size and mass of the universe, speed of rotation of galaxies, stars, and the planet.

Every morning, we witness sunrise and sunset, with tonnes of water rising up and falling back in the form of rain. Every day, billions of seeds sprout from the soil, creating fruits and vegetables that nourish billions of lives all over the world. We are totally unaware of all these processes. These processes maintain the survival of living things and protect our Earth from the known and unknown dangers of life.

Just think about our neighbouring planets—high temperatures of 475 degrees in one region, freezing cold in other areas, waterless regions, and encountering storms and volcanoes. Just think how different other planets are by missing one or two conditions of Earth, how this difference makes other planets dead and silent collections of matter. Our Earth is under sheer protection all the time. The very second you are reading this book, our top layer of Earth's atmosphere is being bombarded by meteorites.

Although we are unaware of this fact, our Earth's atmosphere is protecting our planet every second from terrible disasters. Meteorites from space hit our Earth's atmosphere every second at speeds of 40 km/sec; when they enter the atmosphere, they get burnt up due to the effect of friction. Our atmosphere works like a natural shield for us, protecting us every second. Research studies show that around 50,000 giant meteorites are neutralised by the atmosphere every year. Just imagine what would happen if a single meteorite managed to fall to Earth. The consequences would be catastrophic. The consequences of this single explosion would be 1000 times greater than Hiroshima and Nagasaki.

The vastness of space around you is always restless. Every second, a lot more is happening around you and for you. When you wake up and open your eyes, everything in the universe is set up in harmony and perfectly ordered for your survival and comfort. While you are reading this book, your respiratory system enables you to breathe, your digestive system digests all varieties of food, your eyelids protect your eyes from foreign objects, and your heart beats 70 times a minute—it started its journey of beating when you were in your mother's womb and continues throughout your life.

You breathe approximately 21,600 times a day; without our intervention of thinking about breathing, our body system calculates itself how much breath we need when we walk, run, read, or sleep and makes our lungs work accordingly. Our body's defence system, under our skin, works for us constantly without allowing any foreign particles into our body. Everything is happening even when we are sleeping. Every second, we are protected from surrounding dangers.

A single second of malfunctioning can pose a major threat to our human life, so don't take your time for granted. If you are alive and breathing, then behind this, the Almighty has created countless processes in balance with perfect design and harmony. Does it sound like our lifespan does not hold any meaning? Not at all. Just ponder over the things which the Almighty has set for us. We don't have any reason to depress ourselves, since we have ample opportunities to be grateful for. It is our primary responsibility to reflect upon it. Time is a mysterious entity; you may be fascinated with time travel stories, but it is highly impossible. Why? Because, in order to time travel, you must travel at the speed of light. We cannot regain any of the time we have lost. This life is a one-time opportunity, and we need to use it wisely.

There is a practical way to deeply realise the value of time—visit your nearby graveyards once in a while. It may sound odd, but it is a place of wisdom. Just think, people lying under the soil, once they were walking like us, having fun with their friends, working hard for their families, planning their lives in great detail—unaware that one day they would have to leave everything behind. Perhaps they might have thought this life would last as long as we are thinking right now. But life is full of uncertainties. The time duration in this world passes so quickly than we imagine. We are all living at an appointed time. Every second of our life, we are moving closer to death.

Many people complain about the speed of time because they could not do everything they had planned, especially when you ask older people—they will say their time passed by in the blink of an eye. Most of us are aware of the importance of time, but we fail to grasp the wisdom behind it. At the same time, we should not be opportunistic and run behind worldly benefits throughout our lifespan. Don't try to squeeze all your material affairs into this short span of time. We should be productive and balanced in all ways, for our families and society as well.

There are a few things that people waste loads of time on:

- Dwelling on past miseries and regretting the past. Release the baggage of your past. If you have made a mistake, correct it. We must learn lessons from the past and move on. If you have sinned, repent for it. The fountain of forgiveness is ever rich. Don't waste time by dwelling too much on past worries; it's a sign of negativity. Unleash the optimist inside you.
- Gossiping, making fun of others, engaging in prolonged discussions of trivial matters, and indulging in political arguments waste our time enormously. We should reinforce ourselves by encouraging and advising good behaviour towards one another. Constantly reminding ourselves of our creator helps us display vital virtues such as patience and gratefulness, which will benefit ourselves and our society.

But, no matter what circumstances I was not carried away with negativities. I anchored my faith firmly. I was very reactive, responsible, maintained a balance between my personal and family life. I was sandwiched between the hospital and my personal stuff. I remember it was evening time, I went to the hospital to meet my Dad. When I was about to enter the reception, a van crossed me with a loud horn. It captured every one of our attentions whoever present in the place. A young man from the van emerged out and about to bring out the patient from the van. The nearby security guard stopped him to descend the patient, he asked them to wait until the chief doctor allows them.

While they were having a conversation, doctors were rushed towards the patient. I saw the patient was old age suffering from breathing difficulties. The doctor gave him first aid and they covered the patient's face with an oxygen mask. Then they started talking with the patient's family members regarding the health issue of the patient as well as the admission formalities of the hospital. From my personal experience, I knew that hospital treatment is quite expensive. Post discussion with the doctors, the family members were clueless. I felt they are not affordable enough to admit to the same hospital. After a quick discussion, within a few minutes, they came up with a decision to shift the hospital. They conveyed the same to the doctors. The doctor nodded his head in agreement and immediately went inside. When they were ready to leave the hospital, a labour boy approached them with a slip of paper and unmasked the oxygen from the patient.

The family members were puzzled when they spot paper slips. The paper slip was a bill for the oxygen supply given to the patient for the short span as first aid. Do you think how long the duration of this entire incident? it wouldn't last long more than 15 to 20 minutes. The charge mentioned for the oxygen supply is 3000 rupees for this short span of time. They were shocked including me, but with no other go, they have paid the bill and moved away from the hospital. This incident registered in my mind deeply. It kept running at the back of my mind for a long-time. I consciously inhaled and exhaled the oxygen several times with the feeling of enormous gratitude. Oxygen is a prized possession. Almighty hasn't given you for free without any purpose. If you are alive, then there is a mighty purpose, there is a divine plan behind it. Cling on to hope. Acknowledge your comforts. Enlighten your hardships with your blessings.

You cannot imagine a world without oxygen. If you want to know how it is important in your life, just hold your breath for a few seconds. You will come to know how much we are dependent on it. Every moment of our life, our body requires oxygen to function. Every cell in our body constantly receiving oxygen from the blood. Without oxygen, hardly we can manage five minutes afterward we will lose our consciousness, and eventually, brain death will occur.

Oxygen is a colourless, odourless, tasteless gas. It makes up 21% of the air we breathe. When we breathe, the oxygen enters the lungs and dissolves in the bloodstream. Our blood has a protein called haemoglobin which traps the oxygen molecules and transports them to every cell of the body. Our cells use this oxygen to burn sugar which we have obtained from food. This breaking down of sugar produces the energy our body needs.

For example, if you are burning wood, it uses oxygen and release heat energy and carbon dioxide. But in our process energy released is stored in the cells in the form of chemical energy for our water usage. Carbon dioxide is a waste product, each time we breathe we exhale carbon dioxide. Imagine your body is like a candle, then wax in the candle acts like fuel, it combines with oxygen and produces heat energy. The fuel for our body is nutrients like sugar from our food consumption. These sugar molecules are combined with oxygen to form energy. This energy is used for our daily activities.

When we breathe the oxygen enters into our lungs and gets dissolved into our bloodstream. The solubility property of oxygen is perfectly balanced. If the solubility of the oxygen in water is less, then less oxygen enters into the bloodstream, we would not receive enough oxygen to generate energy. Irrelevant to how much you breathe, the air will fail to reach the cells. It is very tough for human beings to survive with less oxygen support. Without enough oxygen, our body's high metabolism wouldn't happen. If the oxygen is more soluble in water, as a result of this oxygen level rises in blood enormously which leads to oxygen toxicity. It reacts with water in large quantities and produces highly destructive side products. Oxygen is a very dangerous gas if it exceeds the limit.

Our brain constantly in need of oxygen. An adult brain can survive up to 4 minutes without oxygen, a newborn baby can survive up to a maximum of 10 minutes. In order for our brain to receive oxygen continuously three organs in our body are working interactively. When we breathe, air enters our body through the nostrils and to the lungs.

Inside the lungs, oxygen from the air enters the bloodstream. The blood then carries oxygen to every organ of the body through the cardiovascular system. This process occurs continuously without any intervention. As you read this book while sitting in a chair, this process repeats every time you breathe, without pause.

Even a fetus inside the mother's womb receives nourishment through oxygen via the umbilical cord. Their bodies are structured to receive oxygen without the use of lungs. Over time, the lungs of the baby develop, and they begin breathing independently once born. All necessary preparations for the baby's survival in this world are made before birth, highlighting the vital role of oxygen for every organ to function.

The ozone layer in the atmosphere plays a crucial role in protecting our Earth from harmful ultraviolet (UV) rays emitted by the Sun. UV radiation poses significant dangers to living organisms, causing skin cancer, and impairing plant growth. More than 90% of these harmful UV rays are blocked by the ozone layer. Without this protective layer, life on Earth would gradually be destroyed due to exposure to harmful radiation.

Ozone is another form of oxygen (O3), representing an allotrope of oxygen composed of three oxygen atoms. Therefore, the depletion of the ozone layer due to oxygen deficiency would result in the gradual destruction of life on Earth.

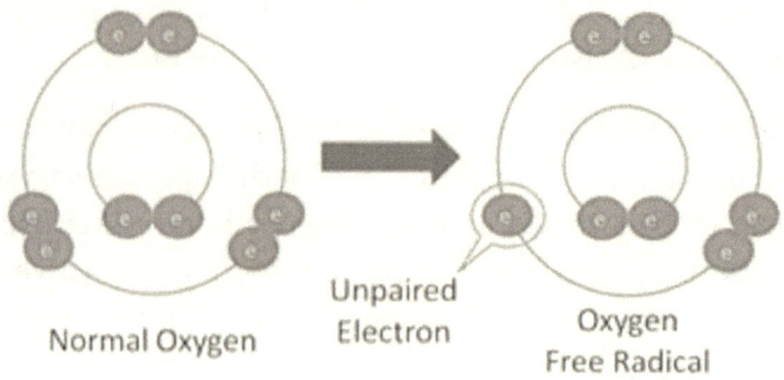

Another shocking aspect of oxygen is that it introduces free radicals into our bodies. Free radicals are highly dangerous and harmful to health. Almost all electrons in the human body exist in pairs. Through various chemical reactions with oxygen, these electron pairs are split, resulting in unpaired electrons, as illustrated in the figure below. These unpaired electrons are known as free radicals.

These energetic unpaired electrons disrupt the function of paired electrons and impede their normal activities. This is why free radicals pose a threat to our health. When we breathe, oxygen enters our bodies and releases harmful substances like free radicals as by-products of chemical reactions. These free radicals are responsible for ageing; they attack the cell nucleus, interfere with DNA information, and disrupt important proteins. This process begins around age 30, accelerates during the 40s, and peaks around age 50, causing the ageing of many cells.

For instance, when you spend extended periods surfing the internet, unwanted information accumulates on your device without your awareness in the form of cache, cookies, etc. Some websites may introduce malware to your device, which can slow down network speed and functionality. We have to format our devices, clear caches, and cookies regularly to keep the system running smoothly. However, unlike our devices, our bodies lack a system to clear these dangerous free radicals on their own. Because life in this world is not eternal, everyone spends a finite period here. On one hand, oxygen sustains life; on the other hand, it inflicts significant damage on the body.

Just imagine, we breathe approximately 23,040 times a day. With every breath, we move closer to death. Regardless of what happens, death is inevitable; we cannot escape it. We live in this world for a predetermined time. Each breath is a lesson; we are here for a divine purpose. Throughout our lives, we should be grateful to our creator, who is in complete control of our journey. If you are feeling low, take a deep breath, release your negative energy along with your exhaled breath. Remember, while you are breathing, someone in the world is taking their final breath. You still have a chance; embrace your trials, conquer your challenges, and emerge victorious.

AN EXHAUSTING DAY

Imagine waking up before sunrise, walking for miles carrying empty vessels, and spending hours collecting water. When you finally arrive at your destination, you see a long line of people waiting to access a deep, muddy hole in the ground. This muddy water is what you must share with your family, community, and livestock. When your turn comes, you hoist water out bucket by bucket using a rope.

As a result, your hands become hard and shredded. Now you have to carry those 20 litres of water all the way back home and serve it to your loved ones. By evening, after completing all your household chores, you have to repeat the water collection process again. It's a busy day without any time to rest. How does this day feel to you? Painfully exhausting, right! Yet, this is the daily routine for people living in regions with extreme water scarcity, such as sandy deserts. Around 10% of the world's population lacks access to clean water. Let's feel fortunate to be among the 90% who do not worry about finding water or the diseases caused by the muddy water that affects the health of our family members. Despite having ample access to clean water, we should not be oblivious to this great blessing from our Lord.

Three-fourths of Earth's surface is covered with water, and 70% of the human body is made up of water. From the tiniest organisms to giant animals, everything depends on water. Water has unique properties that nourish all life on this planet. All our basic needs depend on water. Every drop of water we consume or use for our daily needs is recycled through natural processes. Who knows, you might be using the same water molecules that were once used by the great King Solomon who ruled the entire world. This is because water formed on our planet during the creation of the Earth.

The process of water formation stopped thereafter. Water molecules are formed when hydrogen and oxygen molecules collide at high temperatures and energy levels. Our Earth does not have sufficient temperatures to facilitate this collision. Therefore, no new water can be formed. The amount of water on Earth remains constant; it never changes. The water we drink, and use is the same water that evaporates and returns to us as rain. Imagine if this recycling process did not occur in nature, we would run out of water. Without water, life is not possible. Thanks to the water cycle process, we receive water from rain in a purified form.

Water has many astonishing properties that play a vital role in our planet's life. Let's consider the viscosity property of water. Viscosity is a fluid's resistance to flow. For example, honey has lower flowability than water because honey's viscosity is higher than that of water. Water is one hundred times more fluid than olive oil.

Water has significantly lower viscosity compared to other fluids. The viscous property of water is in a perfect equilibrium state. If the viscosity of water were even slightly lower, the intricate molecular structures of microorganisms would be disrupted, and cell structures would not survive. Conversely, if the viscosity of water were slightly higher, it would be impossible for our capillary vessels to carry blood. More than 95% of our blood consists of water molecules. Imagine if water had a viscosity higher than that of honey; our heart would be unable to pump blood throughout our body.

Oxygen and nutrients from outside enter our bodies; our blood carries oxygen and nutrients to countless cells and removes waste material from our bodies. Our heart acts like a pump, creating flow within the body; our blood travels through the channels of veins and arteries to every cell. If water had a viscosity similar to oil, even if the heart could pump it, it would be unable to travel through veins and arteries throughout our body. The main purpose of our capillary network is to supply oxygen, nutrients, and other substances like hormones to every cell in the body. For a cell to receive this delivery, it should not be further than 50 microns from blood vessels. One micron is one-thousandth of a millimetre; imagine how small this distance must be. If the distance were slightly more, the cell would not be fed with oxygen and nutrients, eventually leading to its death.

This is why every part of our body is completely covered with capillary networks. The diameter of these blood vessels is just 3 to 5 microns, which means 3 to 5 thousandths of a millimetre. If you were to place 10,000 blood vessel capillaries side by side, their combined width would not exceed the width of a pencil tip. The only property that allows water to pass through the blood vessels in our body without any blockage or slowdown is its low viscosity. Therefore, it is highly evident that water has been created with a perfect viscosity level suitable for sustaining life.

Water acts as a universal solvent because almost all chemical substances can dissolve in water. This property of water serves an extraordinary purpose in life. Many important minerals and substances are locked beneath the land, which dissolve in water and get transported to the seas by rivers. Research studies show that every year, five billion tonnes of such matter are carried to the sea, which is vital for sea life.

Before we discuss another important aspect of water, close this book, and go for a run of ten kilometres. What will happen? You will start sweating, right? What's good about that? Let's see. The human body maintains a temperature of 37 degrees Celsius under normal conditions. This temperature is critical; it must be kept constant. If this temperature were to drop by a few degrees, our body's vital functions would fail to perform.

If the temperature rises, we become ill, and the consequences can be severe. A consistently high temperature can eventually lead to death. Our body maintains a critical equilibrium with slight temperature variations, but it is challenging to sustain this equilibrium without certain special properties of water.

Imagine engaging in physical activities from the moment you wake up. Each time you exert yourself, your body generates heat, much like an engine does while running. This heat must not disrupt the temperature balance within our bodies. Two properties of water help maintain our body's constant temperature. The first is water's thermal conductivity, which means it requires a significant amount of heat to raise water's temperature. Since 70% of our body is composed of water, it does not heat up quickly.

It takes a lot more heat to raise the temperature of water compared to other liquids. For instance, an activity that causes a 10-degree rise in temperature in your body would result in a 20-degree rise if your body were composed of alcohol. Therefore, water's thermal capacity prevents our bodies from overheating.

Even a small increase in temperature affects our bodies, and this is where the second property of water becomes crucial. When we expend more energy, more heat is generated, which needs to be dissipated from our bodies. Otherwise, after running just one kilometre, we would risk collapsing into a coma. Our bodies employ sweating as a mechanism to cool ourselves down. When we sweat, water spreads over our skin, keeping our bodies cool and maintaining normal body temperature. These water properties are specifically designed to regulate the temperature of complex life forms like humans.

You may have noticed that there are many polar regions where ice floats on the surface and liquid water exists underneath. Have you ever wondered why ice floats instead of sinking? Two unique properties of water make this possible. Unlike other liquids, which freeze from the bottom upwards, water freezes from the top downwards. This causes the top surface of lakes and seas to freeze in low-temperature regions. If water were to freeze from the bottom upwards like other liquids, continuous freezing from the bottom would eventually lead to complete ice formation with no liquid water remaining. Consequently, life forms would not be possible under the sea or in rivers in low-temperature regions.

Water also has another extraordinary property that allows ice to float on the surface. When the temperature drops, like all other liquids, water becomes denser. However, at a specific temperature of 4 degrees Celsius, water begins to expand and become lighter, eventually forming ice. Water at 4 degrees Celsius sinks due to its increased density, while water above 4 degrees Celsius remains lighter and rises. Only the surface water reaches 0 degrees Celsius and freezes into ice. Because the surface remains ice, it prevents the water below from freezing further.

This makes the water under the ice remain in a liquid state which is sufficient for all the animals and plants in the regions to survive. There are many places in the world has the temperature falls below 0 degrees Celsius, the water bodies become colder and began to freeze, if the water did not behave this way, in other words, the ice did not float all the water freezes into ice form sink to the bottom. This process will continue until all water in the places become ice, then there would be no liquid water left. As a result, much of our planet's water locked up in ice form, which gradually makes our planet lifeless.

But fortunately, that is not the case. Here the million-dollar question is why water misbehaves at certain temperatures from other liquids? Why it suddenly begins to expand at 4 degrees Celsius rather than contracts like other liquids? This is the question, nobody able to answer. Even with the advanced today's technology, scientists could not produce the same liquid with all the above properties of water. Water is the greatest blessing from our creator. Millions of people across the globe don't have access to clean water and get effects by water-borne diseases every moment. Drinking water is not a habit, it's a gift. It has a significant impact on leading our precious life by sustaining our precious bodies continuously. Although it is available for free, we should not take it for granted. Feel gratitude, cherish every sip of it.

MOONWALK

Many moons ago, I was walking in the road heading to my room. I was feeble, dead tired, with an unclear head. There was no one around me. I was all alone but in the company of a beautiful Moon. I was moving forward with baby steps. I need to walk nearly two kilometres to reach my destination. My mind was full of prayers to get rid of my miserable situation. If my prayers were answered by God, then there would not be another night, and I would be walking like this. Yes, this scenario is not happening once in a blue Moon rather it is my usual routine of every nightfall.

For the past two years, this is the still picture of my night. I was working a shift; I start my day by afternoon and works till midnight. From my office, I have transportation till a certain distance, afterward I need to take a walk of two kilometres to reach my room. For me, 2 A.M. is the time to hit the bed. Some nights, it will prolong and extends up to sunrise due to a tight work schedule. Meanwhile, I will miss my Moon companion. Due to shift work, most of the night, I had disrupted sleep. Despite appealing several requests to my management, I couldn't get regular shifts.

My sleep cycle was troublesome. I was facing many health issues, expend much of my money on treatment, which makes my fat purse become lean even at the beginning of the month. I was deprived of sleep, which stands as the biggest barrier to my balanced life. My prayers were to attain balance in all areas of my life, settle myself in ideal sleep hours. I know it's a long-time prayer. But delays don't mean denials. Pain is essential for growth. I have embraced the pain. Nothing good comes easy. One fine day, my prayers were answered, everything around me turned out in a pleasant manner. I have been moved to new team with a raised position at regular shifts.

You can label it as a promotion. Gradually my health becomes normal. I have enjoyed learning new things with utmost happiness. My profession turns into my passion. I used to have quality time with my family members. On top of everything, I am hitting the pillow at the right time without hanging out with my Moon pal. It was an overwhelming moment. I have felt the long wait, made much sense. Remembering the years of pain, Now I am reclining the pillow with my utmost gratitude. Sleep is a blessing from our Lord. The reality of our sleep is still a mystery. Biologists still cannot explain it, doctors and scientists are involved in the continuous study of sleep. Sleep is a precious thing, which everyone experiences every day at no cost.

Sleep is not just a period of rest, rather it is a renewal period of your body. Every one of us spends one-third of our life in sleep. Sleep is one of the basic needs of our body like water, oxygen, etc. Refreshing your mind and your body is very important for a balanced life. For example, how restarting our mobiles, formatting our computer improves the device performances. In the same way, our brain, body parts everything gets renewed while we sleep.

Let's see what exactly happens when we sleep. Our skin must discard 10 grams of dead cells every day, in order to stay smooth. For replacing these dead cells, the same number of new cells must produce in the body. During sleep, our growth hormone will increase. Night hours are the perfect time for increasing our growth hormone. This is because during night hours neither sunlight nor the wind or bodily movements can disrupt cell division. Sleep hours vary from individual to individual, but the ideal sleep duration is 6- 8 hours a day. There are two distinct sleep types: REM (Rapid eye movement) and Non-REM sleep. When you hit the bed Non-REM sleep starts it lasts from 40 – 80 minutes. Basically, a Non-REM sleep slows down everything. When your head reached the pillow, your brain activity slows, thinking slows, your breathing slows, everything slows down in your body.

Then the second type of sleep, REM sleep starts. REM sleep activates your brain, your brain is as active as when you are awake. This is what creating dreams, and it sends signals to your body to act out for this dream. But your body is paralysed like applied brakes on it due to the Non-REM sleep. During this active REM sleep, our body consolidates brain memory like doing filing. So, quiet sleep takes care of our body fatigue and active sleep takes care of the consolidation of our memory.

Sleepwalking is one of the common sleep disorders. The person is half awake and half asleep. When a person sleepwalks, he/she actually looks awake. You cannot tell the difference. The next day if you ask them, they cannot recollect, and they will not be remembering anything. It is common in children. I remember when I was a kid, I used to sleep in the hall, a small distance from the doorstep. One night, I woke up suddenly in sleep, opened the door myself. Sleepwalking throughout our street and reached my grandpa's house residing in the next street.

I started knocking on the door while asleep. It was midnight, my grandpa opened the door and shocked to see me. He was asking me the reason why I went there? But I was still, standing without speaking anything. He comes to know, I came by sleep, without any conscience. He informed my home and laid me in bed. The next morning, he explained to me everything. I was like without remembering anything. Like this, there are many such incidents that happened to me. But some sleepwalking disorder stories are very dangerous and deadly.

In 1987, a man left his bed. He drove his car 14 miles along the highway. Entered the house with a key he already took it along with him, went inside the bedroom. He choked his father-in-law into the unconscious and stabbed his mother-in-law to death. He got back to his car and drove to the police station. When the policeman is on duty, arrived at the station and noticed something strange that the person was sound asleep. He is the evident victim of this sleepwalking disorder. The reason for this sleepwalking disorder might have come from stress or family history but still, we don't have a solid answer for what makes an individual carry out this kind of horrific activity in sleep. But Thanks to the Almighty, during our REM sleep or dream sleep our bodies are become paralysed as a safeguard mechanism.

For centuries, sleep deprivation used as a form of torture. Due to lack of sleep, the victim will get confused. In order to combat it, the victim deliberately injures his body. So that that pain, makes him to hold on something real. But this will not work for long, gradually the victim will surrender and confess. We all of us have an inbuilt mechanism within our body, which controls our daily sleeping cycle. This mechanism is called the circadian rhythm, a biological clock that controls our sleep-wake cycle. But due to some inside and outside factors, this rhythm gets disturbed, and sleeplessness happens.

Insomnia, in other words, Sleeplessness is the most common sleep disorder in our society. Insomnia is a common and curable disease. If not cured, it will lead to serious health issues even death. The majority of road accidents occur due to people falling asleep while driving. Sleep deprivation is the root cause of many serious accidents like Chernobyl, the Bhopal gas tragedy, etc...

We are having a great familiarity with sleep. When we go to sleep, we are certain that we will wake up the next morning, we set an alarm, make plans for the next day. There are many unresolved mysteries present in the sleep. There is no guarantee that we will wake up the next morning. Apnoea is one of the shocking sleep disorders, which is the cessation of breathing during our sleep. It may lead to long term and cause serious life-threatening health problems. There are two cases of Apnoea; In the first case, the brain could not send signals to breathing muscles which helps to breathe. The other case is air stuck in the respiratory tract. Due to Apnoea, oxygen density will get a decrease in the blood. In a long term, it causes hypertension, heart diseases, paralysis. In another way, it causes sudden death while sleeping.

As we have seen from all the above processes, waking up from sleep is a kind of resurrection from death. When we sleep, our life is under so many threats. Sleep is a kind of small death and death is a long sleep. There is a famous quote from one of the pious men of Islamic history; 'If you doubt death, then do not sleep. You will die as you sleep. If you doubt the resurrection, then do not wake up, you will resurrect as you wake up.' Waking up healthy every morning is a great blessing; we should be grateful for our Almighty.

Good sleep is the greatest blessing. Every day is a new day, every morning is a new beginning, every awake is a new life. You are here only because God wants you to be here. Some prayers will take a long-time to get answered. But do your best, He hears our prayers, sees our tears. When you least expect it, doors will open, breakthroughs will come. Your wait won't be in vain. Move forward with complete trust and firm faith. With all my love, Have a blessed sleep!

STORY OF AN OLD WOMAN

This is the true story of an old woman. I have heard it from a renowned Islamic scholar. I can't stop myself sharing with all of you. When night fell, the old woman was alone in her house and praying. It was raining cats and dogs outside. Knocks from her doorstep catch her attention. Puzzled by the unusual knocks at the unusual time, she has approached the doorstep. When she opened the door, she found a man with gentle attire standing in front. He started narrating the reason behind his visit at this odd time; he said, "I was flying from this city to this city, all of sudden clouds were built, thunders and lightning appeared frequently.

Unexpectedly lightning struck our plane, one of our engines got struck. We were predisposed in a situation to land immediately, in order to save our lives. We landed safely in an immediate small airport; since it was a remote area, we found no one to repair our planet. I have approached the pilot and asked, how long it would take to repair it." He says, "It will take a quite long-time, we are just three hours away from our destination, why don't you take a car and reach the destination yourself. I said, "Okay. It sounds good." I drove the car, it was a downpour, within a few kilometres of my travel. My car got stuck in the muddy land. I was unable to proceed further with the car.

I spotted this house from a long distance; without any choice I have decided to knock on. I am in need of your favour if you permit, I will have some food and stay here tonight. It will be a great help to me during this miserable situation." The old woman gracefully invited him inside. The man took off his bag, performed his night prayer, had a hearty meal served by the woman. While he was about to sleep, he sensed that there is someone along with the woman present in the house. He went inside the house, and he noticed a kid lying on the bed, the woman caressing the kid's head by enchanting the words of prayers. The man went nearby the old woman and enquiring about the kid, she said "This kid is an orphan, he is unwell for a long-time. I am the grandmother of this kid. I tried with many doctors, nothing worked out.

The doctors told me that there is one specialist who can help this child. I tried reaching the particular doctor, but since he is one of the top doctors in his field, I got his appointment after six months, it's too far. From that day onwards, I am praying to God. He alone can make my situation easy for me." The man asked, "What is the doctor's name? Soon, he heard the name of the doctor, and he burst out into tears," said the grandma. "It is because of your prayers thunder came, it is because of your prayers lightning struck our plane, it is because of your prayers this downpour happened and trapped my vehicle wheel underground. I am the very doctor you booked an appointment with after six months. God answered your prayers. He dragged me all the way to your home only by answering your prayers." Upon hearing this both bursts into tears, heavily moved by the love from the Lord of the worlds. The doctor took care of the kid with enormous gratitude. Faith prevails beyond limits. Faith world's most luxurious thing available free of cost.

There are many types of research in recent days proved that religious moral values are having a more positive influence on human health. It shows that people of firm faith live a longer and happier life. They are far more patient and become optimistic about adverse conditions. They are able to withstand the problems of our daily lives in a better way. Compassion, forgiveness, and gratefulness are qualities strongly associated with religious people, practicing these qualities linked to decreased stress and enhanced resilience. People who are forgiving others, enjoying physically and mentally healthy lives. According to research, not forgiving others causes negative emotions, which are very harmful both psychologically and physically.

People who have the moral quality of forgiveness gain in two ways. One is reducing the stress from the state of unforgiveness such as anger, bitterness, hatred, fear of being hurt or humiliated again. These psychological consequences lead to physical harm like increased blood pressure, immune suppression, and impairment of neurological function and memory. We have seen people carrying hatred for the long term with their blood relations. Shockingly it goes even after generations. This kind of long-term unresolved anger resets our body's internal thermostat system. If you get used to a low level of anger all the time, you won't recognise what is normal and what is abnormal. It creates some kind of adrenaline rush in our body which we get used to it, it turns out our body and making us difficult to think clearly.

Anger is a harmful state of mind that damages human health in a serious manner. There are many surveys conducted by psychological institutions that showed that depression, stress, and mental illness are seen less in religious people. When practicing religious people see evil in something, they repel the evil with good. Forgiveness reduces anger, pain, stress, and depression. It also leads to positive mental states such as hope, patience, and self-confidence. If not for the pleasure of our Lord, we must learn to forgive people for our own goodness. We should not hold on to anger, hatred, resentment for the long term. Life is too short for such drama. Keep forgiving others and just move on. A lot of good awaiting on your way.

The best gift you can give to your loved ones is to pray for them in secret. Never give up on that. Don't lose hope. Constantly pray for their turning points in life. Prayer works in mysterious ways. Post-2000, ample research was conducted to find out the connection between prayer and health. The study explained people who believe in God are morally and psychologically stronger; due to this, they are able to fight the sickness and recover more rapidly and quickly. There are many types of research and studies that are carried out to study the effect of intercessory prayers, which means praying for another patient's wellness.

A study was conducted between two groups of people, around 450 patients, at San Francisco University. Among those, prayers were read to 150 patients regularly. The outcome of the prayers is amazing. The patients upon whom prayers were read by the unknown people healed faster than others. This is just a single study that I have emphasised; there are numerous research studies conducted to estimate the effect of prayer on health. Prayer doesn't mean oral prayer. It must be active prayer. For example, if a person is sick, he must go to the hospital and take treatment for his illness. After doing everything in his capacity, he must put his trust in God and seek his help. The person must acknowledge his self-helplessness in front of the Lord.

Likewise, if your close to heart people are unwell, after taking care of him/her with all your physical and moral support, you must rely on your creator. We wholeheartedly wish everyone to live a healthy and happy life along with our families. But the world we are living in is a place of test, our life is filled with trials and tribulations. There are good days, and there are tough days. Don't let the difficult days destroy you, and don't let the good days distract you. Never forget the purpose of our life. Be grateful for the Almighty. If you see someone ill, pray for their well-being. Prayer is your beacon of hope.

Gratitude is the primary key to happiness. People who are practicing gratitude experience more positive emotions, feel more alive and express more kindness and compassion. Science has proven the amazing effects of gratitude in people's lives. Masaru Emoto was a Japanese scientist. He studied the molecular structure of water. During his lifetime, he has conducted extensive research on the molecular structure of water. He has proved that human consciousness has an effect on the molecular structure of water. His experiment with water crystals showed, how the attitude of gratitude transforms physical matter.

As a result of his experiment, he has released groundbreaking images of water crystals when they were exposed to different emotions. If the water molecules are exposed to the thoughts of gratitude, words of prayers, any other good words, or good consciousness. It has a positive influence on water. The shape of the water molecular structure gets changed. The molecular structure formation is fascinating, pleasing delightful to see. On the other hand, if the water is exposed to negative thoughts, emotions, or bad words, the molecular structure becomes disfigured and unpleasant. His research further proved how polluted water after being exposed to prayers, its molecular structure is restored and becomes clean and healthy water.

The following image gives you a better visual understanding of this experiment. Just think: our body is 70% made of water, three-fourths of our Earth's space is occupied by water. Water is a living consciousness. By practicing gratitude, we can gift a positive impact on our bodies and the world around us. Faith is the most necessity for our happy life, how rainwater replenishes the Earth, and nutrients give energy to our body, in the same way, nourishment for our soul, is faith. Faith is deep-rooted, it goes beyond hope. Faith is accepting all the ups and downs in our life respectfully, knowing that the divine decree of the Almighty drives everything. Even at the sheer difficulties which tore us apart, faith has the power to make us whole again. Faith is a powerful tool to a successful journey of our life available free of cost. You can achieve anything with firm faith. You can push yourself beyond boundaries. Never give up, There is always light at the end of the tunnel.

THREE

Lucia

One holiday morning, I was surfing my mobile phone. The display picture of my father-in-law grabbed my attention. It was a young age photo of him with western attire. In reality, he dresses quite opposite, wearing our cultural dresses like dhoti. I guess it is an occasionally captured photo, maybe in his marriage ceremony. I have found something much different in that particular image.

It was the colour of the photo. A black and white photograph. It reminisces my childhood days. Those days, hardly one or two houses has a television in a street. The entire street kids including me crowded together in their houses to watch a movie which is broadcasting once a week, Sunday evening. Sometimes I used to sleep there while watching, my parents have to search for me a lot and find out me. The television runs with a single channel. Everything running on the screen with the colours of black and white. You cannot predict which colour dress, the hero or the heroine of the film is wearing. My uncle was a foreign return, own a phone with him.

When he is at home, he used to lend his phone to me for 10 minutes every evening. I remember I was super happy; my adrenaline rush out of happiness when I grab the phone from his hand. I enjoy those 10 minutes by playing games of catching eggs, space impact, and snakes. Those black and white displays registered deeply in my brain. It was a time, the world of technology has not yet entered into the vivid world, the world of colours. Colours, painting our world. A world without colour is a great void of fun. It brings delight to our souls.

Imagine you are living in a world without colour. Try to visualise everything in the world lack of all colours except black and white. Your body, your pets, fruits, flowers, sky, oceans everything is in black and white. How do you feel, how boring the sight would be? Nobody ever wants to live in such a world. Even black and white itself comes under colours; we cannot imagine a world of colourlessness. What if all of a sudden, our world loses all colours? Then everything will get mixed with each other. It is difficult for us to distinguish between the objects. You cannot differentiate between flowers, fruits, fluids, objects, etc... It will be quite an irritable situation for every one of us. For example, you are going to your friend's house for the first time.

In the street, every house followed a uniform pattern of architecture. How will you keep your friend's house in your memory? It is hard to keep things in mind without colours. It plays a vital role in man's communication with the outside world. It helps us to establish a connection between people, places, and events as we remember everything by their external appearances. The colourful world appeals to the human soul. But the story behind these beautiful colours is far more complex and fascinating than they appear. All living creature survives and thrives only in the presence of colours. The only being on Earth, which can give thought to the understanding behind nature is humans. Let's have a bath in the world of colours with its astonishing journey and fascinating details.

It's a long journey, only light can make it at light's speed. Hold hands with me and thrill yourself with the thriving journey of colours. The story of colours begins with light. If there is no light, no colour. The Sun is the main source of light and heat for our planet Earth. The Sun is one among billions of medium-sized stars present in the universe. There is no life on Earth without the Sun. Even if a single aspect of the Sun changes, it would unimaginably affect our planet. The energy emitted by the Sun reaches our Earth in the form of waves. Imagine you are throwing a pebble on the pond, how the waves are formed in different lengths and different dimensions.

In the same way, the energy emitted by our nearest star, the Sun, has different wavelengths and different temperatures. Each star in the universe emits light of different kinds, different temperatures, and different wavelengths. The light from space can be classified according to its wavelengths. The wavelength of light ranges from gamma rays, which have the shortest wavelength to radio waves which have the longest wavelength. This is what we called the spectrum of colours. Among the wide range of the spectrum, sunlight emits light of wavelength, which is restricted to a very narrow range of the spectrum which includes ultraviolet rays, visible light, and infrared rays. When do scientists start studying why the Sun concentrates only this narrow range of emitting light among the wide variety of spectrum? they come up with an interesting conclusion that these rays alone are required to maintain life and colour on Earth. Life on Earth would not be possible without these minuscule band of spectrum.

Visible light is the only reason, we can see things around the world. In the spectrum, above visible light is the ultraviolet rays, and below is the infrared rays. Ultraviolet violet rays reach the Earth in a very minimum amount. Our atmosphere filters the entering of ultraviolet rays in a large amount because it has damaging effects on living organisms on the Earth. The ultraviolet rays reach the Earth in large amounts; they damage the tissue of the organisms and cause death in large numbers. If the amount of UV rays is very low, the required energy of living beings would not be fulfilled.

Our Earth's atmosphere filters the UV rays and sends them to Earth in an optimum amount. Infrared rays reach the Earth in the form of heat; they pass through the atmosphere and provide heat to the Earth's species. With the help of visible lights alone, our eyes can see the images and colours of the world. Humans and animals make use of this light to make the objects visible.

The light from the Sun crosses its first phase in space and reaches our Earth's atmosphere. There are several gases present in the atmosphere, which reflect all the harmful gases from space and allow only visible light and a small portion of infrared light. It blocks all the harmful radiations from space, like x-rays, gamma rays, and ultraviolet rays, from reaching the Earth. The properties of the gases present in the atmosphere have the ability to block all the harmful radiation from space from reaching the Earth.

At the same time, these gases are permeable to visible light and a small portion of infrared rays, which is vital for the survival of life on Earth. The densities of these light rays in space are very strong, but when they reach the atmosphere, due to the strong gases present in the atmosphere, they make the strong light rays from space diffuse and weaken. Because of these properties of the gases present in the atmosphere, we can directly see sunlight without any harmful rays or strong light rays attacking our eyes.

The light passes the atmosphere and strikes the Earth's object at a speed of 300,000 km per second. With this enormous speed, the light strikes the atoms present inside the object. The atoms of the objects reflect the light in different wavelengths, corresponding to the different wavelengths different colours are produced. For example, we are all familiar with Newton's prism experiment. White rays of sunlight comprise seven colours, which we call VIBGYOR.

When the white light passes through the prism, it splits the visible light into seven different wavelengths, which gives rise to the birth of seven different colours. Likewise, if the atoms of a particular object reflect the visible light in the range of red colour wavelength, you will see the colour of the object as red; if it falls into the wavelength range of blue colour, you will perceive the object in blue colour. In the same way, the book you are holding right now, the lines you are seeing, your room, trees, and all the outside world reflect the light of their own wavelength.

The colour of the object mainly depends on the pigment molecules present inside the object. Objects of different pigment molecules produce different colours. This pigment molecule present in our eyes, skin, plants, and all the outer surface of the objects. There is certain energy needed to activate these pigment molecules. To activate the pigment molecules, this energy level should neither be more nor lesser than the optimum. Surprisingly enough, our visible light has the exact energy level to activate these pigment molecules present in the objects belongs to the Earth.

For example, Chlorophyll is the pigment molecules present in the plants. When sunlight strikes the plants, the chlorophyll present inside the plants reflects the light in a particular wavelength corresponding to the colour of green. This is the reason we can see green colour in plants. Chlorophyll is the most dominant pigment molecules present in the plant, which is why most parts of the plants look green in colour. Apart from this, the plant receives energy from the sunlight and produces carbohydrates, which is the primary source of food for all living things.

There are numerous pigments present in nature. Every pigment gives a different colour of corresponding wavelengths to the light. Melanin is the pigment which gives colours to our eye, skin, and hair. Based on the proportion of the presence of melanin, our eye colour varies from black to brown. Melanin also serves as protection to our eyes from harmful rays. It absorbs higher energy light more strongly than the lower-energy light. It absorbs more strongly the ultraviolet rays than the blue colour of lesser energy.

Melanin pigment apart from producing colours gives optimum protection to our eyes from hazardous rays. Our blood contains pigment haemoglobin which gives us a red colour. Pigment molecules vary from different species. For example, the blood of a snail is blue in colour because of hemocyanin pigment present in the blood, the blood of a cockroach is colourless due to lack of pigments in the blood. Pigment molecules are the reason for the colour diversity of our vibrant world.

Still, the journey of light is not yet completed. The reflected light from the objects must reach your eyes, then only it can be perceived as colours. Our eyes convert these light rays into nerve impulses that reach the vision centre of the brain. The human eye is a very complex structure. The light from the object reaches the retina of our eyes. Our eyes have cone cells of three types, each response to three different colours of blue, green, and red.

These three colours are primary colours. The rest of all the other colours are formed by the combination of these primary colours. We have short cone cells and long cone cells. Short cone cells help us to see the colours and long cone cells help us to see shapes and movements. The cone cells are sensitive to these three primary colours at different degrees which forms the combination of millions of colours. These cone cells convert the collected colour information into nerve impulses. The nerve cells attached to the cones carry these impulses to the vision centre of the brain.

This is the last stage in the journey of light. The nerve impulses convey the information to the vision centre of our brain. Here is an astonishing thing happens. Our brain inside the skull, a piece of meat in a dark place decipher the information of colours and produces a

colourful world inside our brain. Whatever colours, images you are seeing, everything is produced in the vision centre of our brain. A

matchless miracle of creation is the human brain. Scientists still unravel the countless mysteries of the human brain. We have no control over any of the above stages of colour formation. It's a great blessing from our creator for the delight of mankind.

Colours play a very important role in the life of living beings. Every living being must know the meaning of colours to survive. Colours help animals to recognise their mates. They help to absorb or diffuse heat from the skin of the animal. Using these colours, living things protect themselves from enemies. If they did not use colours to defend themselves, they would be easy prey to their enemies, and soon the entire species would become extinct. Camouflage is the most defensive technique used by animals to hunt their prey or hide from their enemies. Animals adapt their skin colour according to their surroundings so that it is hard to get noticed by their predators. A frog can adapt the same colour as the underground, a butterfly adapts the colour same as the tree bark, a snake takes the colour of the tree branch and gets mixed with it to catch the prey. Camouflage is a special technique used by many living creatures.

This picture shows the camouflage technique exhibited by a toad according to its floor colour. Some animals use these colours as a symbol of warning or other means of communication. For example, flowers which need pollination for reproduction attract insects using their vibrant colours.

The fox hides inside the grass to catch prey, using its skin colour to blend with the surrounding environment.

Male peacocks use their coloured feathers to attract female peacocks. Some other animals use colours to protect themselves from heat and cold. Animals living in cold regions change their skin to black colour so that it absorbs more heat energy and keeps their bodies warm. This is achieved using special enzymes secreted inside the animal's body. In summer, when temperatures are warmer, they change their fur colours to white to keep their bodies cool from external heat.

Birds with multi-coloured feathers exhibit a diverse variety of colours. The reason behind this diversity is the pigment named keratin present inside the feathers. Some of the colours exhibited in bird feathers are because of the keratin pigment. Another interesting aspect is the inner structure of their feathers, which is similar to the structure of a prism. When light enters the feathers, due to this structure, light reflects and refracts, giving rise to colours of different wavelengths that are more vibrant than the colours produced by pigment molecules. Due to environmental conditions, these feathers shed frequently and renew gradually. Each time a feather is renewed, it exhibits the same colour.

Everything in the world has its own unique colour. Every colour has its own meaning and information to convey. There are animals with striped colours on the skin like zebra, tiger, cats, etc. None of the two zebras has the same stripes. Each zebra has its unique stripes like the fingerprint of a human. More than beauty colours are highly important for the survival of living things. There are countless blessings God has bestowed upon mankind. If we start counting the blessings of our creator, we could not enumerate them.

Adorned with Accolades

The Super Power Author Award of 2023 was presented to the author of the book "Embrace Magnificence" by Adhyan Books during an event in Delhi.

Lavishly honored by the dynamic philanthropist and revered president of TIME Matric School Mr. Kaja Shareef in the author's native land for the work "Embrace Magnificence."

Revered by his alma mater's prestigious Golden Jubilee Matriculation school celebration for the work "Embrace Magnificence" at the grand English Literary Association ceremony of 2023.

Acclaimed with the prestigious 'Achiever Award' by the distinguished Aram Awards, a respected social welfare foundation in Tamil Nadu held at Madurai, for the literary achievement "Embrace Magnificence" in the year 2023.

FOUR

Ying and Yang

Yin and yang are like both sides of the same coin. Its symbol represents the duality or polarity in nature. It is more than 3000 years old Chinese philosophy. It is the concept to see both the good and bad things of any entity. It says about the harmonious balance between two contrasting and diametrically opposite forces which you can see everywhere in the universe. Our whole universe is composed of opposite forces, these forces rely on one another for their existence. One thing alone cannot exist without the other. You cannot feel the pleasure of the shadow without experience the heat of the scorching Sun. If the yin force is stronger, then the yang will be weaker, and it is vice versa.

God alone is the sole divine entity, the rest of everything that comprises the side universe exists in pairs. God has created everything in pairs. For example, Male and female, light and dark, heat and cold, love and hate, black and white, even plants have been created in pairs. Recently we have discovered that plants have gender distinctions. Still, many unrevealing things in the universe exists in pairs. Scientists undergoing many groundbreaking types of research to uncover the concepts of duality in every entity of the universe. So that they can extract the harmonious nature of the particular entity. Because It is only possible to harness the complete benefit, power, or energy of a particular object when it is in harmony. I still remember, in my 12th standard physics book, the seventh chapter completely speaking about the dual nature of light.

It seemed a very interesting concept to me, out of curiosity I studied and memorised the complete chapter without leaving a single paragraph. The fun part is I had a fluorescent colour highlighter, I started highlighting important points, but I ended up with highlighted the complete chapter. It was my favourite chapter in my 12th standard physics. The nature of yin and yang is kind of negative and positive. In reality, you cannot pinpoint relatively like, this is yin, and this is yang. Because one thing seems yin to you, the same thing is perceived as yang for others. For example, If you are standing on the roof of a tall building, relative to you the roof is yang since it is way taller than you, it elevates you.

But relative to the sky, the roof is considered as yin. It's like yin is down and yang is up. If you look at the below image, yin is the dark colour swirl and yang is the light one. Another important concept behind this symbol is that it is not necessary for yin and yang to be two different entities. Each side contains the seed of the opposite side. You can see the black dot in the white region and the white dot in the black region. It means each object has its own positive and negative nature within itself. For example, You can use a sharp knife either harmful or helpful to others, the internet you are handling every day under your fingertips you can make it beneficial by bringing the world under your feet or you can make it disastrous by indulging in lustful activities.

Human beings are created with the gift of free will. It is every individual choice to exercise good or bad deeds. So that they can cultivate their souls and reap the same with either yin or yang. To bring harmony to anything, it is necessary to study the pros and cons of the particular being object. Yin and yang is a universal concept. You can apply it to any creation or entity. In the following pages, we are going to discuss the two sides of a living being, which is one of the oldest life forms on Earth. Even right now, at this moment while you are reading this book, you are at the company of those beings. They are our lifetime companion. They are seeing you in a way while you cannot see them. They are studying you, exploring you all the time. They know you more than you know about yourselves. They are helpful and harmful, devotional, and dangerous, a friend and a foe, your well-wishers, and Ill-wishers. Welcome to the world of bacteria.

When a baby is born, it is fresh and free from bacteria. As soon as the baby arrives at the world, within hours more than hundreds of species of bacteria get colonised into the organs of the newborn baby. It starts living in their mouth, tongue, intestines, etc... If you start counting the number of bacteria in your mouth alone, the number of bacteria will exceed the number of people living on our planet. We are destined to live with bacteria, throughout our lifetime. There is no other way. They keep us healthy.

Our body is made up of trillions of cells, but bacteria are single-cell organisms. It cannot be seen by the naked eye. You can think, you are reading this book alone in your living room. But you are accompanied by a large number of bacteria within you and surrounds you. Bacteria can survive in any type of harsh environment. They can even live under the depth of the sea where water temperature exceeds 400 degrees Celsius. Bacteria live in every nook and corner of the world. There are different types and species of bacteria; some of them depend on oxygen for living, some of them live without oxygen, some prepare food themselves using photosynthesis, some prepare food by breaking organic substances, some live on the Earth's surface, and others live underground.

It has come in different forms with different functionalities. Bacteria is a single-cell organism that has a cell membrane. The bacteria have different functionalities. Because of their minute size of less than 0.005 mm. It is very difficult to study the inner structures of the bacteria. Bacteria have a long hair-like structure called a flagellum. It helps bacteria with locomotion. It has a molecular structure different from the cilia and cell membrane. It is the only organ in the world which moves both forward and backwards. The cell membrane of the bacteria is more powerful than that of human skin. Because of this strong, resistant nature of the cell membrane, it can adapt to very high or low temperatures. It can live inside the human body, under the soil, float through the air, under the oceans, and even in a toxic chemical environment. Some of them even resist radiations.

If the conditions are favourable, by cell division a single bacterium divides its number by two within 30 minutes. This two becomes four, this four becomes eight. This reproduction continues and it reaches millions of bacteria within 12 hours. There are some species of bacteria unaffected by any high or low temperatures, some other species even withstand radiation of more than 1000 times fatal to the human body. If the environment is hard to live in or not favourable, by the time these bacteria undergoing an astonishingly shocking process, to survive in the harsh environments.

Based on the environment, bacteria take different forms to thrive. Under suitable conditions, it reproduces by giving two offspring. These two are identical in all the features, which further reproduce, and the bacteria started spreading and surviving in the environment. But if the environmental conditions are not suitable, bacteria thought it is very difficult to get nourishment or to live in a particular environment, in that case, it gives birth to two offspring. But this time, these two are not identical. One is larger than the other cell. The reason behind this inequality is, only one cell is going to survive. The main cell or the larger cell bacteria absorbs its own sibling, as a protector.

For more than 8 hours, it will use all the energy for nourishment from its sibling. The absorbed cell forms a strong protein sheath in the body of the large cell bacteria. The smaller bacteria dies after it gives all its nourishment and formed a protein sheath in the larger bacteria. The primary motive of this small bacteria is to take care of its sibling, by sacrificing its own life to form a protein sheath. This protective sheath becomes more resistant. This highly resistant structure is called spores. This process of formation of spores is called sporulation. The bacterium with spore starts to survive in the environment by reproducing their offspring. This is the reason, why bacteria are more difficult to destroy.

In 2019, our world faced a global tragedy. Amazon, a biodiverse forest, burnt at a rate this world has never seen in a decade. It was burning for more than a month. Humans across the globe are frightened as it disturbs the ecological balance on a large scale. In YouTube, news about the Amazon forest fire becomes more viral in all the diverse channels of science, environment, general knowledge, etc. They have captioned it as the Amazon forest- The lungs of the planet Earth. I have found that there is a discrepancy in the particular caption. I agree it is a disaster; this incident affects our ecological balance, and it supplies a large amount of oxygen to living beings on Earth.

But the Earth's largest portion of the oxygen is produced by microorganisms. Photosynthesis is a must process on Earth to sustain life on Earth. Humans and animals lack the mechanism of direct use of solar energy. They obtain solar energy in ready-made form as a result of photosynthesis processed by plants and microorganisms in the world. The bacteria species, known as Cyanobacteria, generates more than 50% of the Earth's oxygen. The mechanism used by Cyanobacteria is similar to that of plants. The majority of these bacteria contain chlorophyll. It uses sunlight to produce energy.

This energy is stored in the form of sugar. It produces billions of tonnes of sugar and oxygen by the process of photosynthesis. This sugar and oxygen are used as essential things for the survival of other living beings on Earth. Despite their smaller size, their numbers are unimaginably great. It exists over a larger part of our planet. The energy produced by photosynthesis by these bacteria is of greater importance for all the other living beings on Earth.

Like how much we have dependent on oxygen, we need nitrogen to survive. The structure of proteins, vitamins has nitrogen. Nitrogen gas is present in the atmosphere at about 78%. But living things cannot use nitrogen from the atmosphere directly. It is needed to turn into a usable form. This process is fulfilled with the help of bacteria. There are various ways nitrogen reaches our Earth's surface. Below are a few:

- During lightning bolt, due to rapid heating or cooling the nitrogen gas combined with oxygen, which forms nitrogen oxides. Nitrogen oxides on the Earth's surface are converted to nitrates with the help of Nitrate bacteria. This form is absorbed and used by plants.

- There are some bacteria present in the roots of plants like pea, beans, etc... absorbs the nitrogen from the air to the soil. Bacteria provide nitrogen to the plants, which is very essential for the plant to grow. Plants supply sufficient nutrients to the bacteria to survive. Just imagine, we are getting nitrogen only by consuming plants and animals which takes nitrogen from plants, without this bacteria-plants partnership, our survival would not happen. This is one of the world's best partnerships for our planet's survival.

- When the animals and plants die, nitrogen present in the remains of the dead plants or animals processed by bacteria. It converts the nitrogen into Ammonia. Ammonia is further converted to nitrates. It is used by life living plants.

In the same way how atmospheric nitrogen gets converted into nitrates by bacteria which in turn used by plants; there are some denitrifying bacteria present in the soil which converts the ammonia or the used nitrogen from the animals or plant remains into nitrates. Nitrates are again converted by another denitrifying bacteria into atmospheric nitrogen. So, it gives back the nitrogen to the atmosphere. Thus the nitrogen cycle is completed. Every living creature on the Earth directly or indirectly depends on photosynthesis. Nitrogen is a very important element for photosynthesis. Bacteria play a main role in the life of all living beings by turning atmospheric nitrogen into a usable form by plants and animals.

Personally, I must thank bacteria in one way because it making my kid's favourite food, Yoghurt. Yes, bacteria is a good chef. Most delicious food which ornaments our dining table made from bacteria. There are some bacteria which does not breathe oxygen, it produces his own energy by breaking down the sugar in food products. While it generates energy using sugar molecules, the products released by the reaction make the food more delicious. Like the carbon dioxide released during the reaction puffs the bread.

This process is called fermentation. Cheese, yoghurt, wine are formed by the process of bacterial fermentation. During the process of fermentation, bacteria release vitamins and minerals which is exceedingly good for our health. It maintains our body's metabolism. Doctors recommend the fermentation food for cholesterol problems, the reason behind this is, these microorganisms help us to regulate cholesterol level in our body. The bacteria produces food for us in an amazing way, it is highly beneficial for our health.

There are bacteria in the underground which are helpful to isolate gold, the most valuable commodity on Earth. The normal bacteria divide once in 3 to 4 hours. But bacteria that live underground divide once in 100 years. The lifespan of underground bacteria lives a million years. They feed on rocks and isolate the gold particles beneath the surface. It helps us to refine gold. The bacteria present everywhere in the world, there is no place in our body where the bacteria is not present.

Escherichia coli an important bacteria lives in our intestine which feeds from our food sugar particles and in return it helps us for digestion and vitamin absorption. Likewise, it helps animals to digest their food. Animal body not able to digest the cellulose present in the leaves of the plants. Bacteria help them to digest the cellulose of the plants. Despite their smaller size, bacteria play an extraordinary role in the survival of the world.,

So far, we have seen the good face of the bacteria, but there is always a dark side of the Moon. There is always the yin nature of yang. Yin and yang will flow together. There are many ways bacteria lead to the death of living things. For example, if you enter the supermarket, you have seen loads of bacteria protection products like health soaps, hand wash, sanitisers, antiseptics, etc. Because a single bacterium even less than the size of one micrometre is enough to cause major harm and kill a human being. Harmful bacteria enters into humans or animals bodies through food.

We have already seen, how bacteria divide and multiply once it finds a suitable environment. The nutrients present in the food appropriate for the bacteria to multiply in large numbers and survive in the environment. It begins to reproduce in the food and starts secreting a poisonous toxin in the food. When we consume food, the organism begins to reproduce in our intestines. Wherever the toxins present, it causes the death of the cell in the particular region. These harmful activities lead to food poison and other harmful health effects.

In Europe, during the year 1347-1351, nearly 200 million people died as a result of a black pandemic named the plague. The bacteria named Coccobacillus, which is the prime cause of this disease. This bacterium lives as a parasite in animals like rats and squirrels. It enters the human body through the eyes, mouth, respiratory passage, etc... when this bacterium enters into the human body, our body shows huge resistance, but gradually it is unable to withstand it for a long-time. The toxins secreted by these bacteria enter our lymph glands.

Due to the entry of the bacterial toxins, lymph glands swell up. As a result, it causes the tissues in our body to die and begins to decay. Gradually, the bacteria enters into the bloodstream, reaches all the lymph glands of our body, causes the lymph glands to swell which failed other organs and death occurs. All these disasters begin from a single-cell. It is not seen by our naked eye. It can be only observed with the help of a microscope. Vaccines discovered by humans end up as failures against the intelligence of these bacteria. Because the plague bacteria is having genes that are naturally having resistant to different antibiotics.

Even if we come up with new vaccines also, it starts developing its own immunity and creating the problem much worsen to the human body. It is a bitter fact that we were truly powerless in front of these invisible creatures.

When my kid was about 2 years old, accidentally he has consumed a piece of sponge. It happened in front of my wife. By the time I was away from my family due to work. But he is good and active. But my wife was restless, she is hyper care about our kid's well-being all the time. She was frightened, completely obsessed with worry because of this incident. She called me and started worrying a lot. I couldn't convince her with my consoling words. But my kid was active, with no symptoms of vomiting or any other sickness. Then I told her that, "Please wait until he is passing stool or faeces. If he does so, it will expel out the waste, no need to worry." It sounds rational to her. Fortunately, within a few minutes, my kid passed stool, she was convinced a bit. Even after, for her console, I have arranged a call with the doctor.

After a conversation with the doctor, she was more relaxed. The reason, I am sharing this is, our body has a powerful digestive mechanism. Gastric acid secreted inside the stomach break and digest the foodstuffs. It is very powerful enough to dissolve razor blades. But, have you ever thought, how this powerful acid present in our body without destroying our stomach? But bacteria knows our body and it functions more than ourselves. For years we have believed that peptic ulcer or stomach ulcer is caused by food and stress. But, In 1982 we have discovered that the main cause of peptic ulcers is a bacteria named Helicobacter pylori or H. pylori. It is capable of making a shelter in this dangerous environment.

Our stomach has a mucous layer, to prevent damage caused by its own acid. The bacteria H. Pylori hides inside this layer to protect from the acid. It starts affecting a particular region of the mucous layer and hides under the layer. Once it finds a place, it starts secreting toxins. The main purpose of this toxin is to kill the immune cells in a particular region. After the generated toxins killed the immune cells of the location, now the place is suitable for the bacteria to live. It thrives and starts reproducing to spread the disease. Once, the mucous layer of the stomach gets damaged, our body sends a large number of immune cells and nutrients to the damaged location.

It is not happening in all the other regions. Especially to prevent the mucous layer damage, our body sends a lot of nutrients. Our body continuously sends the nutrients until the damage in the mucous layer gets healed. Bacteria present in the mucous layer, keep on feeding the nourishment, and starts spreading the disease. We can prepare a large list of diseases, these single-cell organism causes. But finding the remedy for the disease caused by these organisms is highly difficult.

Socrates, a famous Greek philosopher sentenced to death by consuming poison himself. Throughout centuries, poison death is a notable plot to kill the person. Because the reason behind death would not be found easily. Only in recent years, we have advanced enough in medical science to find out the poisoning death through Autopsy. With the advancement of emerging technology on one positive side, humans also empowering their strategy in a parallel window on negative. Yin and yang should flow together, Right!

During the times of World War 1 and World War 2, there are many nations used biological toxins like bacteria, viruses to kill humans and animals as a strategy of war. Bio war which causes human life comes to an end. Using biological weapons, we can devastate the enemies on a global scale. These living organisms as biological weapons are highly resilient, nonpredictable, very tough to destroy. Terrorists started using biological weapons because it is cheap to manufacture than missiles, but it can create huge harm. During this bioterrorism attack, scientists believe that the widely used microorganism is the bacteria called Bacillus anthracis, which causes Anthrax. Anthrax causing bacteria produces spores.

It survives a long-time, very cheap to manufacture in a laboratory at the same time it can cause highly potential harm to the people. The Anthrax attack can happen in many ways; Anthrax spores can be released into the air using a plane, whoever breathes the air, will get infected with Anthrax. On September 18, 2001, an anonymous letter was mailed to the media office located in Washington D.C. The United States. This is the first letter in history that contains Anthrax spores. As a result of the letter, five people died inhaling Anthrax and 17 people get infected. This was a highly dreadful incident and one of the most complex cases in history.

Duality is part of nature. There are good and bad, light, and dark, angel and devil, life and death, heaven, and hell. But there is a significant difference between humans and other beings in wisdom, intellect, and free will which means the ability to distinguish between right from wrong. If we don't have free will, why we are here? There is no purpose for our existence. God has created us with an inbuilt security mechanism inside our heart, Guilty conscience. It will alarm you when you go astray or involve in evil deeds. It alarms us when the first time we get involved in sins.

We should not take it lightly or consider the minor sins insignificantly. Ignoring the divine caution, when we keep on indulging in the sins, it will eventually pile up become a mountain. We will be used to it; we will start enjoying the sins without even having a little guilt. Because we have already suppressed our inner caution. Like a famous proverb 'first crime everything, second crime something, third crime nothing.' It eventually overtakes us, destroys our soul and life. So, Release your negative energy. Infuse your soul with positivity, love, and peace. Be grateful to your creator. Live and let live. The choice is yours to be a yin or yang, light or dark. Begin your venture towards success, not a failure, joy not misery, heaven not hell. Even if we have committed so many bad deeds, it's not too late to continue our life with good deeds. Indeed Good deeds erase bad deeds. Spread positive vibes. Be a light.

FIVE

You Are a Masterpiece

It's been more than four centuries, since human started this journey to wonderland. During this journey we have witnessed countless wonders. No matter how far we have travelled, the road seems endless, complicated, and filled with fascinations. This journey unravels the mysteries of mankind, which seemed impossible at one-time. Beauty is but skin deep. Travelling under your skin is a journey to wonderland. Understanding the miracles of our body make us feel special to be alive. It will add more meaning to our life. At this moment, trillions of cells are working at your service. Everyone is specialised on their tasks.

They are engineers, chemists, army men carrying versatile knowledge in every activity. Your eye cells processing each letter you are reading right now, your stomach digesting the food you ate few hours ago, your immune system continuously waging war against the foreign bodies which are trying to enter into your body, billions of tiny powerhouses to power the whole cellular world. All these processes taking place at an extraordinary speed. None of the cells refuse to accept their tasks. Each cells doing their job in a perfect manner. Discovery of electron microscope paved the way to reveal a lot about the invisible universe under our skin. I have tried to throw light on very few sections under our skin.

You will fall into love and admiration when you start studying these flawless activities. Humans exist between all the wonders between the universe and the micro world. Among all the creations, which

Almighty scattered throughout the space, he purposefully honoured human beings in all the aspects. Each one of us is special. Man is the masterpiece of God's creation. A masterpiece is the one which has no duplicates. Imagine if there is any other creation superior to human, it would have dominated the mankind long-time back. A masterpiece is extremely valuable.

No matter how much far a man can deviate from divine path, the fountain of forgiveness is rich. The door of repentance is always open until your last breath. A masterpiece is cherished by the artist more than other arts. Almighty created man by his own hands. He loves us all the time beyond our imagination. Each one of us are special and unique. Man is the masterpiece of God.

THE MAN INSIDE A MAN

We humans are given with the gift of intelligence. Every moment our world is evolving. Every day breakthroughs are happening. We are in a continuous state of improving ourselves. Our world is moving towards with a series of advancements in all arenas. Close your eyes, think deeply, and try to extract the realistic picture from all parts of our today's living world. Billions of human populations across the globe. Everyone is unique in their own way. Everyone has a plan; everyone has their own targets to achieve. People of different regions has different complexion. People of diverse regions around the globe, communicates using thousands of languages.

Digital communication engages with the people of hundred miles away. Reaching people at any parts of the world is just a click away. Advanced transportation technologies which increases our mobility. Roads, rails, airways, and giant cargoes sailing across the seas. We are able to reach any parts of the world with ease. Life built on energy, by which we light our homes, transport goods, power our devices for almost all things we rely on it. Mountain of information stored on the clouds. We can store, copy, or edit any number of data, access it at anywhere, anytime. Strong fighting force of army men, across the borders of every nation.

They should always be vigilant, withstanding themselves even under rough weather and protecting the nation against foreign invaders. This is how the picture of our modern world come into picture on our minds. Could you think any of the above phenomenon without taking human into account? No it's not possible. For the sake of imagination, even if you keep all the above processes without human intervention, it would not be meaningful at all. But you can reimagine everything by replacing an unseen microscopic living cells instead of human. A single living cell does everything what a man can do except the freedom of choice. Yes, it make plans by its own brain of nucleus, communicates with other cells, transports enzymes, hormones without even a small delay.

It harnesses energy by our food and generates power throughout our body. It stores million pages of data inside its DNA. It's immune cells all the time fighting against viruses and foreign particles with its extraordinary defensive mechanism. Exploring the unimaginably smaller things like cell structure, which seemed impossible for a long-time. Modern scientific inventions of electron microscopes made it possible. Universe, once a mystery for decades. But, now the more we learn about the universe, it seems simpler. On the contrary, the more we study about the cell, it becomes more complex even in today's advanced age of science. Inside you, there are trillions of men working for your well-being constantly without any rest. This is the story of the man inside a man.

Our body is made up of trillions of cells. Cells are the building blocks of human body. Each cell in our body, lives, communicates, generates power, fights, reproduces, and dies in a world invisible to our eyes. These cells are very small. If you gather a million of cells in a single place, it would not be larger than a pinpoint. Then Imagine how small the size would be. But with the invention of powerful microscopes, this world no longer out of sight. World within our body is a world of cells. The first cell of a human body produced in our mother's womb by the union of one cell from our mother's body and other one from our father's.

This single-cell starts dividing and forms new cells continuously which becomes our flesh. It forms different specialised kinds of cells like skin cells, brain cells, blood cells etc... we have more than 200 different types of cells in our bodies. Each type has its own specialised functions. For example, blood cells in our body transports oxygen throughout our body by means of blood vessels, muscle cells help us for our body movements like walk, run, jump, lift weights etc... skin cells protects our body from microbes and other foreign objects entering into our body.

Likewise different types of cells has their own different shapes, functions and goals to achieve. Each cell has cell membrane, which is a semi permeable membrane. It acts as a skin of the cell. Cell membrane identify things which is coming inside and going out. It allows things which the cell needs and keep the harmful substance out. The cells recognise one another, for example brain cells recognise brain cells, blood cells recognise blood cells. Every cell knows where to go, which organ it should form, by which degree it should multiply, when to divide and when to stop. Cells belongs to a particular organ, join together and make up the same organ by recognising each other.

Cells in our body communicate each other in an astonishing speed. The cells in our brain linked to the muscle cells throughout our body. With the help of nerve cells, our brain is constantly communicating with all parts of our body. It is sending and receiving signals to every part through nerve cells. This communication happens by means of electrical stimulus transmitted along the network of nerve cells in your body. If you want to flex your muscles, your brain sends electrical signals to your arm muscle cells through the nerve cells with the information of how much degree it should bend, on which speed it should flex like that, receiving the information our cell in the arm exhibits the same behaviour. This communication happens with an amazing speed of less than one-thousandth of a second. It occurs constantly without any time interval.

Our brain constantly sends and receives electric stimuli from all parts of the body. Even at this moment, when you are reading this book, nerve cells at your fingertips sends message to your brain regarding the weight of the book, your brain receives and sends stimuli to your muscle cells to exhibit a lifting force suit with the weight of the book. At the same time, your brain sends stimuli to your eyes, ears, feet, and the other parts of the body as well. This communication works even without your intervention, say if you step foot on a sharp object, without having a thought you withdraw your feet by means of reflex, involuntary responses. This is a security mechanism for the protection of our bodies.

There are thousands of hormones, which serves as a messenger molecule. It allows continuous communication among cells, makes critical decisions, stimulates the needed secretions at right time and right quantity. Absence of these hormones, our body would not be in an organised state. For example, Growth hormone regulates our bone growth, insulin hormone regulates amount of glucose in our blood, oxytocin hormone initiates secretion of breast milk during childbirth etc... One cell can perceive only one chemical message from a hormone.

The cell and hormone message should harmony with each other like a lock and key. A right key can open the lock, so for the other cells this message is nothing to them. Therefore the message from a hormone not mistakenly set some of the other cell into action. The moment message reaches the cell, forwarded to the nucleus of the cell which set the cells into action by receiving this message. Every cell in our body has its own advanced communication systems.

Each cell has nucleus at its centre, which is the brain of the cell. Inside the nucleus DNA of the cell is found. During the time of cell division, the DNA found inside the cell should be copied for the new cell. DNA acts like a databank, it stores huge amount of information concerning particular living organisms. A single DNA contains billions of letters. If you want to write down every information in a single DNA, you need to write an encyclopaedia of 1000 volumes with a million pages each. But, during cell division copying or replicating this DNA for the new cell will not take less than an hour. This means, million pages of writings can be copied within an hour without any single mistake.

When we communicate to people using English language, the 26 letters of the English language are the base. From that we use words, then it combines into sentences. Likewise, in the language of DNA the base letters are A, T, G and C which are called Nucleotides. From these four base letters, all the information regarding our body, eyebrows, skin colour everything will be present. The order inside the DNA is also very important. Just think millions of letters written in a jumbled manner, would it make a meaningful article? No. In the same way every information inside DNA placed it in a particular order. This information is getting copied while cell division without any single error. Even machines with the advanced technologies for copying will not meet the same.

We are wondering about architecture of human wonders of Eiffel tower, Taj Mahal etc... our body cells constructs human skeletal system with its impeccable functionalities. It can withstand for any type of stress and strain produced by the environmental conditions. It helps us to maintain our body strength throughout our lifetime. There are 206 bones in our body, which are interconnected with each other. Some are soft, some are harder than steel, each bone location is very delicate and intelligently planned. Joints are present at the connections points of our bone. With the help of joints, we can easily bend our back, run, jog, pickup any objects by means of joints. For example, If you are having a car, the door should get connected to the car in a rigidised or fixed way.

If it is so, we cannot open or close the door. It serves no purpose of the door. The door should be connected with revolute joints so that it can rotate in one axis direction, this rotation helps us to open or close the door. Likewise each connection part of the bone, installed with specified joints which does it particular function of the organ. The outside of the bone is solid, while the inside of the bone is porous honeycomb structure which makes the bone light weight. If the bone is completely solid without this porous structure. It would be very heavy to lift. At the same time, it is not fragile, due to the solid outer surface it has multiple times than the more strength than the structure of steel material. Scientists inspired from the bone structure and replicate it in different material. They were able to increase the material fatigue life 100 times more than their actual structure. It paved the way too many monumental architectures on Earth.

Energy is very important for our daily activities. We, humans spending so much money for extracting energy from various sources like water, wind, Sun by building dams, windmills, solar panels, nuclear power plants etc... we are spending much money on generating the power. Energy is vital and very expensive. Let's see how the world of cells generates power and keep us active all the time. Energy we need to live, generated in the power houses inside our cell called Mitochondria. Mitochondria is made up of proteins. The nutrients present in our food is converted into a useful form of energy by serious of chemical reactions.

It is stored in our body in the form of tiny packets called ATP molecules, which is readily available energy packets. All the process inside our body takes place by using this ATP molecules. ATP molecules, (Adenosine tri phosphate) is called the energy currency of the cell. Oxygen plays a very important role in production of energy. Every day our body uses 48 kilograms of ATP. But amazingly not a single point of time, our body's ATP molecules amount exceeds one gram. This is because, ATP molecules are not stored in our body. The readily available energy packets prepared for our immediate use of energy. The life of our cells is completely dependent on this energy. Every second, trillions of cells present in our body, use billions of ATP molecules for energy production. When a cell needs energy, the ATP molecules breaks and release energy which is used by the functions of the cell.

This happens at amazing speed. ATP or Adenosine Tri Phosphate molecule has three phosphate groups; when one linkage of phosphate is broken off, energy is released, and ADP (Adenosine Di phosphate) is formed. If two phosphate group broken off, then AMP is formed (Adenosine Mono Phosphate). From AMP, with the release of the last phosphate group, ATP molecules all energy gets transferred into the relevant molecules. Some of our body cells do lots of work like muscle cells because of frequent contraction and relaxation. These cells have more Mitochondria than average cells to produce more energy. Mitochondria which is less than the size of a millimetre, but the power generation capacity of these cells are way more complex than today's man-made nuclear power plant.

There are cells inside our body continuously protecting our bodies fighting against enemies, bacteria, and viruses like a well-disciplined army of a nation. When a foreign particle enters into our body, the immune cells will not simply start fighting with them in the first go. It begins with identifying the nature of the enemy and obtain the intelligence. Based on the intelligence, it starts the war against the enemies. If it had not gather the intelligence of the enemy, our immune system starts attacking our own cells by mistake. The first cells which come in war against the enemies are phagocytes. These are the front line of defences. It engages in face-to-face contact with the enemy.

But sometimes phagocytes could not match with the speed of the enemy, the enemy cells multiply very quickly. In that case Macrophages cells comes into action. It attacks enemy groups in masses. It secretes a protein, which sounds the general alarm in the body by raising its temperature which causes fever. Macrophages plays important role in destruction of bacteria and other harmful viruses. Macrophages has another important property, which traps and engulf virus. Then breaking it apart a particular section of virus and carries it like a flag. This part will act as a sign for other immune cells and use the intelligence for advance planning in future.

All the moment you are reading this book, your cells are constantly involved in activities for your well-being. Some are copying your DNA, some are generating energy for all parts of your body, some cells are transporting information between the brain and the other respective organs in our body, some are waging war against foreign particles. Whatever we have discussed so far about cells is not even a drop of an ocean. Still scientists unravelling the mysteries of the unseen world. Inside our body, completely a different world, made of cells. They are working ceaselessly for our well-being without our intervention. A sign of flawless creation from our creator. Studying about this is a gratifying experience. Like it or not, we have a solid dependency on our creator. He bestows us all the time. Let's our life propel forward in faith.

JOURNEY LOST

Long-time back when I was a child. We were going to a family trip. When we have reached our tourist spot, Courtallam. It is the spa of South India. It has many health resorts, ancient temples; but out of all, the prime tourist attraction which makes this place special is numerous waterfalls and rivers. Breathtaking visionaries, mesmerising beauty of this place soaks your soul with delight. Season times, this place is always overcrowded. When we are heading towards the waterfalls, there are many roadside stalls on both sides of the way.

It is pretty much fun to watch and shop. My younger brother, he was at the age of 6, distracted by the plenty of showcased toys and eventually lost us. In a short time, We have realised ourselves that we have lost him and started shouting his name and searched him everywhere amidst the crowd. He was also realised that being lost, he kept on wandering and looking for us with a grip of agony. Few minutes later, he composed his nervousness, stopped moving further. He was very carefully looking at the crowd for the sight of his parents and listening whether he can hear it out our voice or not.

Due to overcrowd, he was unable to spot us or heard our voice either. As a smart kid, he traced back his pathways and reached the van, we travelled. After a long-time hunt, we found him nearby our van when we returned. Every one of us overjoyed by his sight.

This is common with every one of us, getting lost in our childhood days. Even though we are more familiar with our mother or father voices, we could not recognise it, when we lost. It is due to the external noises around us, which make us incapable of pay attention to it. Similar way, these worldly affairs tend to distract us in all the possible ways. We are getting lost in this materialistic world which drags us to detach our divine attachment. Every one of us here, are created with inbuilt divine guidance. But we are keep on getting distracted ourselves a lot by indulging into this worldly pleasures and pains.

We need to apply brake and try to listen our inner voice. He has given us intuition. He has given us the freedom of choices. We are accountable for our life's shortcomings. Still his guidance is within us all the time. It's up to every individual to exercise restraint and pay heed to the divine guidance within him. He never abandon us. If you give life to your inner voice, it will give life to you. You have guided every moment by your Lord, even before you were born. If you want to know how? Embark with me on this marvellous journey. Let's rewind everything. This is the beginning of every one of us.

This is an incredible journey of every one of us. When a girl attained puberty, her reproductive cycle starts. Every month women's body produces egg and makes the preparation for the process of fertilisation, which means fusion of sperm and egg cell. At the beginning of every month, pituitary gland in women's brain produces a hormone called LH hormone. These hormones are micro molecules, they dissolved in blood stream and start their journey from the women's brain to the ovaries. Movement of this micro molecules through our body is equivalent to the journey of many kilometres. As if the LH hormone has our body's anatomy map, without deviating to any organ or anywhere in between our body, it reaches the ovary without getting lost anywhere in between the journey. Where does this guidance comes for this hormone? The women's ovary has number of immature egg cells, when LH hormone reaches the ovary, it sets few of the egg cells into action and make it mature.

There are number of hormones in our blood, but none of the hormone does this job except LH hormone. Among all the egg cells in the ovary, only one cell fully matures and released from the ovary. With this prior information, we dive into the journey of male reproductive cell sperm. Body of an average man produces 10,000 sperm cells with each heartbeat. This is our natural state before we have been given this physical body.

Sperm cells, they have their own life. They are free living cells. It's freely moves all around the place, they are constantly in motion. A sperm is just one 500th of a millimetre long. It has only one target in his lifetime, to deliver the genetic information of male to the woman reproductive cell, egg. Interestingly our sperm cell is having both gender information. It has both x and y chromosome. While joining with female egg cell which already have XX chromosome, if the sperm delivers X, it becomes a baby boy, if the sperm delivers Y, it becomes a baby girl. Gender determination chromosome present in the sperm cells. Sperms produced in billions stored on top of the testicles of a man. It has to wait.

On contrary to the billions of sperm cells producing continuously in man, a woman body produces single mature egg cell in a month as we have discussed in the beginning. Then just imagine how the battle will be, reaching the single mature egg cell will become a heavy competition between sperm cells. It's a journey filled with dangers; many sperm cells lose their lives during this journey. The environment is similar to a battle ground with series of complex difficulties. Amidst all the crisis, sperm cells has to make their own way to the target egg cell.

During the time of intercourse, a healthy man ejaculates average of 300-500 million sperm cells into the woman body. Sperm cells are forced into the woman body by muscle contraction of testicles, without involving any free movement. The time man experiencing his utmost pleasure, the sperm cells begins its adventurous journey. Soon it reaches the vagina, millions of the sperm cells came out and die. Everything thing inside vagina working against sperm, it's a matter of seconds between the sperm arrival and it's death. Even a saliva inside the vagina cause the sperm lifeless. Sperm cells face attack in all directions. On top of everything, the acidic environment in the vagina makes more than half of the sperm cells into death. This acidic environment produces by the female immune system, as far as the woman body is concerned, the sperm cells are just a foreign particle. So this acidic environment, try to kill all the invading force. A single second delay of the sperm cell is enough to get doomed.

After lost more than half of their friends, the survivor sperms now ready to reach the cervix, an entrance hole to the uterus. Cervix is way high above them to reach. Imagine your life is dependent on climbing a ladder to reach the sky. In the sperm world, it is trying to reach a kilometres high ladder to reach out the entrance hole of the uterus. It has to go against the flow defying gravity. A single wrong turn will cause the sperm cells to face the death. While this climbing process many of the sperm cells give up and lose their lives. Only 1% of the sperm cells make it to reach the cervix.

Usually cervix is tightly closed, but for few days of every month, it will be open. Now the successful sperm cells enters into the opened cervix. Inside the cervix, the sperm cells continue to swim. It has to exhibit the ideal swimming characteristics, if any of the sperm shows random motion it will get excluded in the journey. The resident cells from immune system already present in the uterus mistaking the sperm as foreign invader and destroys more. Meanwhile the matured egg released from the ovary waiting for the sperm to get fertilise.

Out of hundred million cells, few dozen only comes near the egg. The fast-swimming sperm reaches the egg within half an hour, while others might take days. An average lifespan of sperm is 48-72 hours. The first sperm which come into contact with the egg, will fertilise the egg. After the perilous journey and against incredible odds a single sperm cell fertilise the egg. As soon as the sperm cell enters into the egg, the egg cell membrane releases chemicals which push the other sperms away from the egg and creates an impenetrable egg membrane. Due to this reaction, the egg membrane hardens more and more. So, inside the egg the single sperm cell is tightly packed and outside no other sperm cells get attached to the egg. These unlucky cells gradually lose their lives by not succeeding their purpose. Felicitations to those who are reading this book, you made it. Amidst all the dangerous odds, you have sealed your victory.

After fertilisation, the fertilised egg start growing fast and dividing into many cells to form an embryo. This embryo starts its preparations to settle in the walls of the uterus. The surprising factor here is although the genetic material of the embryo is different from its mother, the immune system cells could not attack the embryo as it attacking the sperm cells, other foreign invaders. It remained as a mystery for a long-time. Recently science unravelled the mystery. Every cell belongs to our body has a universal password. When body's immune system attack the cell, it communicates the password and rescue himself.

The women's immune cells starts attacking and try to eliminate the embryo in the uterus, the embryo cells sends a message which is similar to the universal password of the body cells. Receiving the message from the embryo, the defence cells recognise it as a friend not an enemy and abort the attack, allow the embryo to settle in the uterus. How does the embryo, a cluster of cells, gained this consciousness? We are living in an era of advanced scientific technology, researchers found that among all the 300 million sperm cells, everyone is unique characteristics. Everyone carries different genetic information. If some of the other sperm cell succeeded in the journey, then you would be a different person right now.

Now inside the womb of a mother, the embryo starts its journey to become a baby. It's a completely new world for every one of us from a drop of semen to a grown baby. It's a world of 9 months. We are completely taken care by our creator. The heart begins to beat during the week 5. The brain, spinal cord and other organs beginning to form. At eighth week, the developing baby called fetus. The baby will feel independent of his mother support around the eighth months. Generally it will take 40 weeks to deliver a baby. When the baby step foot in this physical world for the first time, it's starts crying. Because this is a new world to the baby, it was accustomed inside the womb of a mother for 9 months. Now, it is the first time the baby seeing this new world.

The newborn baby nutritional needs met by feeding mother's milk. It is one of the miraculous nourishments created by God. The ingredients of the milk meet all the nutritional needs of the baby. This ingredient will change itself according to the baby needs. The production of this milk is a hidden miracle for years. A special hormone secreted in the women's pituitary gland of brain called prolactin, which activates the milk glands to produce milk. In the beginning of pregnancy, there are number of factors stopping the prolactin hormone to produce milk. There is no point in secretion of milk before baby birth. How is it getting stopped during pregnancy? During pregnancy, the female reproductive hormone oestrogen, surges up in the body of women. This oestrogen hormone induce the secretion of another hormone called PIH (Prolactin Inhibiting Hormone), which in turn halt the secretion of prolactin during the pregnancy period. Imagine you are driving a vehicle in a downhill, how you have constantly apply pressure on brakes?

In the same way our body suppress the secretion of prolactin during pregnancy. Once the baby born, the oestrogen level in the body goes down, which reduces the secretion of PIH hormone. Eventually it lifts the brake on prolactin secretion, which in turn starts the production of milk for baby's nourishment. How does our body cells learnt to stop and secrete the prolactin hormone at the right time?

These are the marvel of creations. Think for a moment, you have guided all along the way to become a human even before you were not able to exercise your freedom of choice. We are never left abandoned at any point of time in our life. No matter wherever you are standing in your life, the light of guidance reaches you like how the Sun rays reaches every nook and corner of our world. In depth of our silence, we have our inner voice of divine guidance.

It can be lost, or you may be forgotten by so many worldly distractions out. Listen to your inner voice. It is a beautiful space where nothing exists, and everything exists. You are not you and I am not me. It is a field where you can harness your divine energy. Give your consciousness to reach out the depths of silence residing inside you. It's pure bliss.

MORTAL COMBAT

Childhood memories are evergreen. It is irreplaceable. Human souls have natural affinity towards children because they are innocent and less corrupted. I remember, I have spent much of my school days in gaming. I used to get pocket money of one rupee every day from my mom. I save each penny and spend for playing video games. There were one or two gaming centres near our locality. For five rupees, shop owner allows me to play for 30 minutes. If we have completed any of the game successfully, we will get a bonus of extra one hour. This kind of offer will boost us a lot to spend much time and money on games.

These are some of the tricks of the trade implemented by the shop owners, to drag more kids. I have played and completed lots of games. One of my favourite games was Mortal combat which means fight for death. It is a two-dimensional game. One of the most popular games in 90s. It is the battle between the fighters. Player must choose any one of many unique characters to fight against the other player. Each character has its unique stunts. We need to press the right combination of buttons to reveal the stunts and defeat the opponent. The game is full of blood and violence. I still remember the finishing touch of the game, once you defeat the opponent, the defeated character was falling into the spiked pits and his body would be bathed in blood.

This kind of scenarios often, we have seen only in virtual. Despite of being a couch potato, I was gaming inside a room. My body is constantly under siege. A huge army is defending its homeland from enemies. Every moment of life, our body is fighting for death. This is the real combat started since billions of years ago. For our body's immune system, cuts and wounds are part of everyday life. They obstruct attackers, destroy invaders, and even repair its own damage. Just think we fall in sick and get recovery many times in our life. What happens inside our body? Our body is a battle ground, our immune system obstructs and attacks the entry of foreign organisms all the time. It is a bitter struggle of every moment of our body.

Right now, at this moment your immune system for your body fighting against unseen enemies. You are reading this book peacefully. Without the help of our immune system, you cannot complete reading this book. Everything around us are unseen monsters. Our planet of human body constantly waging war against them. I have deliberately dedicated this section of my book is to introduce the brave warriors within us. This is the tale of valour.

Bone marrow is the birthplace of our specialised troops. Bone marrow is a spongy tissue present inside our bones. It generates

RBC (Red blood corpuscles) and WBC (White blood corpuscles). Our blood contains all the necessary first aid kits for the infection and it carries oxygen to the cells of all parts of our body. WBC is the formidable army against enemies. These are the cells of immune system, which fights against infection. During the incident of Hiroshima and Nagasaki, atom bombs were dropped. The radiation released by the explosion caused many people die after two weeks by continuous bleeding and infection.

Scientists conducted experiments by animals to explore about what happened due to this casualty. The results were revealed that the radiation killed blood forming cells in the bone marrow. These cells are responsible for blood clotting and immunity. Because of lack of production of these cells, humans were succumbed to death by bleeding and absence of defence against infection.

Our skin is the front line of defence. The serves as a shield to our inside warriors. If it gets cut, our body releases immediately an army to the injured spot for defence and repair the wound quickly. We all might have observed; if a slight wound starts bleeding, without few minutes the bleeding stops. What if the bleeding goes on without stopping, it drains the complete blood inside body and cause the person to die of excessive blood loss. In order to avoid that our body maintains an automatic security system called blood clotting.

When a cut happens on your hand or knee without any delay platelets, colourless blood cells arrives on the spot. It releases chemicals, which makes fibrin. Fibrin is a thread like structure. Fibrin obstructs the flow by weaving a web. It traps the RBC and the platelets together on the wounded region. The platelets stick together to plug the wound. This process forms the blood clot. The weaved web stiffened over time, which is what we called scar. Then the process of healing begins. If you are living in a castle, the skin serves as a strong wall of the entire castle to protect you from the entry of enemy forces.

Another doorway for our enemies is our respiratory tract. There are many unseen microbes are present in the air which surrounds us. They are trying to enter our body through our air passage. But our body setup a barrier within us to avoid gaining entry of the microbes. Our nose secretes a fluid, called mucus. Mucus sweeps out 80 to 90% of the microorganisms entering into our body through nose by dust particles. Sometimes due to more bacterial infection, our nose secretes more fluid. This is what we call running nose or nasal mucus. Another security mechanism inside our nose which contains tiny hairs called cilia.

Cilia produce series of air currents by moving, like the rhythmic movement of paddy in the paddy field when a gentle wind hits it. This movement causes the foreign particles move towards the throat where they are swallowed and later getting discharged by the acid in the stomach. At times, these microbes struck in the windpipe. It trigger our reflexes to cough or sneeze. The coughing action forms a high-pressure air burst, which causes the microbes in the windpipe to expel out at the speed of 960 km per hour. Even formula one race cars could not match this speed. Then Imagine what an astonishing security mechanism installed in our body.

Food is another mode of transport by which microbes can enter into our body. The guards of our body waiting vigilantly at the region where the food finally ends up, the stomach. Once the microbes, overcome all the obstacles and reaches the stomach, it's time for it to counter a new security mechanism. Our stomach produces gastric acid, which is highly acidic. This is an unpleasant situation for all the microbes.

Most of the harmful microbes gets killed by the gastric acid. Even if some microbes managed to overcome this situation due to not getting enough contact with the gastric acid, they must encounter another conflict with another guard located on their way. That is, the digestive enzymes secreted by the small intestine. The microbes which entered through food completely get defeated here. Our fighting guards are present everywhere in the body to protect us.

Breaking all the barriers, if any enemy managed to get inside our body, then that is where the real war begins. Our body started unleashing phagocytes. The first soldiers of our body to meet the enemy. Phagocytes flows continuously within our body and keep our body's situation under control. It eats out the bacteria like scavengers. There are two types of phagocytes, Mobile and Immobile. Mobile phagocytes travels throughout the body and fight against enemies. Immobile phagocytes, which sits at gap between the tissues, fights with the microorganisms at the particular region without moving.

Phagocytes destroys all the antigens or enemies. If it couldn't manage to fight with the enemies due to large number of enemy forces, it raise an alarm throughout our body. But how does it raise an alarm? When the phagocytes gets disintegrated by enemies, the cells burst and forms a pus to overflow. This pus formation activates the lymphocytes, lymph nodes and all the immune forces throughout our body. We will look into about lymphocytes in the following pages. This alarm sets the activation of second wave of defence, which is Macrophages a type of phagocytes.

Macrophages destroy number of enemies at a single time. It has the longer life span of months, some of them lives even years. It absorbs and digest all the foreign particles. If the enemies are more in number than the number of Macrophages, then Macrophages releases a chemical called Pyrogen. It is a kind of alarm. Pyrogen takes a long journey to the brain. It stimulates the brain to increase the temperature of the body, which in turn to become fever.

When the brain developed fever, the person becomes sick and tired. He naturally forced to take rest, now all the energy of the cells completely utilised to defend the enemies. So, the energy will not spend in other activities except in defence. The Macrophages will do another interesting activity. Once it defeat the enemies, it transfer the information about enemies to the all the other immune cells of the body. The defence strategy of the particular microorganisms gets transferred by Macrophages. It will be used for future encounter of the same type of enemy.

Our body is completely surrounded by military camps, army men and military intelligent forces. This is called lymphatic system. Like circulatory system, this system runs throughout our body by means of small tubes called lymph vessels. Lymph nodes are the military camps, present at certain spots of the lymph vessels. This military camp produces army men, called lymphocytes which fights against the infection. The lymph fluid are the military intelligence, which travels through the lymph vessels throughout the body.

While travelling it gets in contact with all the tissues of the body and return to the lymph vessels by carrying an information about the tissue. If any defence or hostile activities of the tissues, present in the information which immediately transferred to nearby lymph nodes. Lymph nodes acts as military camps and starts producing army men, which is lymphocytes to fight against enemies. Lymphocytes plays a very important role in defence system. They check our body cells several times a day, if it find any sick or old cells, they will destroy them. There are 100 trillion cells in our body, lymphocytes present in our body is nearly one trillion.

Compared to our body cells, lymphocytes are only one percentage in count. Yet it does regular health check for all the cells several times a day. Just think during the crisis hours of COVID-19 how each nations suffered to conduct a single test even for suspicious people alone. Then Imagine the job of lymphocytes, it conducts test several times for all the cells in our body each day. Indeed it is a miraculous creation of God.

Lymphocytes present in our body are two types: B cells and T cells. Each has their specialised defence techniques. Let's explore one by one. Lymphocytes or immune cells are produced in the bone marrow. Some of them becomes fully mature and transported to lymphatic tissues which are called B cells. B cells produce weapons against enemies. The weapons are made of proteins are called antibodies, which is harmful to the enemies. It is used to attack the enemy. The antibody has two main purposes; first purpose is to bind the antigen, which is the invader cell and second is to destroy the antigen.

B cells produce different antibodies for each enemy. Because antibodies must be compatible with the enemy structure. Antibodies produced for one disease might not be effective for another disease. It is not an easy process. It is like manufacturing compatible keys for millions of locks. A single B cell produce ten million weapons an hour, which means ten million antibodies an hour.

Imagine if a single B cell produces different antibodies for each million of enemy cells, then how much infinite the wisdom behind the creation. Sometimes these weapons confuse their targets and starts attacking our own body cells. In that kind of scenario, other body cells sends signal to the particular B cell to commit suicide. Receiving this signal, an enzyme secrete inside B cell which decompose the DNA of the cell.

Finally, B cells which produces antibodies harmful to enemies alone stay alive in our body. There are other type of B cells, which are not participating in defence. These cells are called, memory cells, which keeps on tracking the molecular record of the enemies. It has a strong memory. In future, if it encounters the same type of enemy, based on the previous record it stimulates the other B cells to produce respective antibodies. It accelerates the war in a faster and effective way.

How does this B cells affects the enemies? Microbes like bacteria and viruses carries a chemical on their body called antigens.

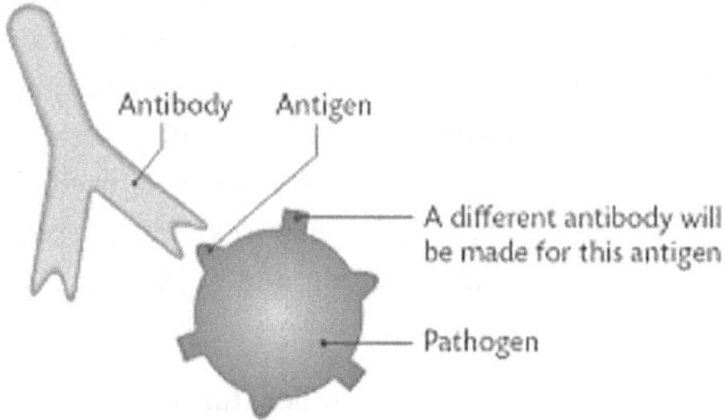

As shown in the image, B cells produce antibodies which fits exactly with the antigen of the enemies. Antibodies binds with the antigen and make the enemy cell ineffective. It inactivate the enemy cell. The enemy cells marked with antibodies, easily spotted as foreign particles. Now the strong defence carried out by the Macrophages and T cells, the enemies would die soon. Some of the lymphocytes produced in the bone marrow, transfer to thymus, which is located in front of the heart.

Lymphocytes multiply and mature in thymus are called T cells. There are two types of T cells: Helper T cells, killer T cells and Suppressor T cells. Helper T cells comes into play at the initial stages of the war. When the Macrophages carries the information regarding the enemies, helper T cells absorbs the information. It also decipher the dyed antigens and gives signals to stimulate killer T cells and B cells. Receiving this signal, B cells starts producing weapons called antibodies respective to the antigen.

Once the bacteria or virus gets inactivates by the antibodies, killer T cells kills the invader cells. Killer T cells secrete a chemical substance called Perforin. It perforates the cell membrane of the target enemy cells. Afterwards leaking starts from the cells and eventually it dies. After the war has won, Suppressor T cells stops the defence activities. The helper T cells and memory B cells alone present in the blood, in order to activate the war if the enemy of same type spotted out again.

Let's go through a brief summary of our immune system, whatever we have discussed so far. Once the bacteria or virus enters into our body, phagocytes starts fighting against foreign particles. If phagocytes could not manage to counter the enemy, Macrophages comes into action. Macrophages fights with large number of enemies at a single time. In addition to that, it collects the information about the enemy cells.

This information gets conveyed to the helper T cells, which studies the nature of antigen and stimulates the B cells to produce compatible antibodies and stimulates the killer T cells to fight against the enemies. When the antibodies binds with antigen and inactivates the enemy, killer T cells kill the enemies by drilling holes in the cell membrane of enemies. After the battle has won Suppressor T cells, suppress the immune system from defence.

Just imagine like the Mortal combat game we have discussed in the beginning, what happens if we are taking control of our own defence against the invading enemies? We would not realise the entry of the enemy cells into our body. We will come to know only when we experience the symptoms of the disease. By the time, the enemy cells would have been settled in our body long-time ago. Even if we start fighting after this, how can we generates weapons, or the antibiotics based on the invading forces. It is difficult to identify the location and the nature of enemy cells, we would be needing a full body check-up.

Even a small cold is enough for us to cease our life. But fortunately we are not in a situation, even to think about this. This explains how we are in dire need of our creator. His infinite knowledge and wisdom encompasses everything between heavens and Earth. He protects us all the time. But Are we mindful of our blessings? If the answer is yes, then there is no room for doubt, even the entire universe wages war against you, by the will of Almighty you will emerge out as triumphant...

THE FIRST MAN

When God begins to create the first man, he took soil from different parts of the Earth. He took it from the rocks of diversified colours; black, white, brown, yellow etc... He took it from the brown soil, red soil etc... He brought soils of different properties and the different colours from different parts of the Earth. He added water into that. He moulded it into clay. God fashioned it into the form of a human being with his own hands. So, now it's just the clay with shape of human form. That is why we are born with different colours from different parts of the world. It is completely hollow inside with no blood vessels, no oxygen.

It is just like a statue with no life. It was an empty, hollow statue. It was placed in the heavens. Angels in the heaven saw this, they want to know about this new creation of God. They are curious about the new being in their neighbourhood. They go nearby the statue, they used to touch him. Since it is hollow inside, when they poke the statue, it creates a tickling noise. So, when an angel touch the statue, they get scared and used to run away from it. They don't know about what's God is going to create a new being. Now, God blow or breathed his soul into the statue. The soul travelled all along the body from head to toe.

Once soul goes inside, it allowed the statue becomes alive. He is Adam, the first man. Adam found himself alive. Then God taught him the names of everything to Adam. God, the knower of seen and the unseen taught mankind directly. After he taught the names of everything to Adam, he called angels. In order to honour mankind, Almighty ordered the angels to prostrate Adam. Obeying God's command all the angels prostrated except Iblis, he became the devil. God asked the names of few things to the angels. But the angels did not know the names. They could not reply. Angels immediately admitted to God by saying, 'We don't have the knowledge of these things except what you have taught us.'

Then Our creator asked Adam to tell those names, Adam replied by answering those names. Adam knew the names of those. Angels were absolutely amazed. Soon they came to know, this is the new being God has created, this is a remarkable creature. They glorified their Lord with his praises. What are those names God taught to Adam? For example, Adam taking a seed, he is putting the seed into the ground. Then the seeds are growing, and he is cultivating it. Then, the tree grows. Adam is taking that tree, cutting it.

He makes the planks of wood out of it. From the planks of wood, he makes something which has four legs, to sit on it as a table or chair. This is something that the angels never knew. If the angels see the chair, they will not know; What it is? What is it made for? They wouldn't know the buildings, they wouldn't know the chariot, they wouldn't know the pyramid. They wouldn't know whatever the technologies mankind going to innovate. They wouldn't know of the new things and the difference between them. Angels have the knowledge of things only what God has given to them. But this is completely new for them. Almighty created humans superior over all the other creations.

Because God planned man to become the viceroy of Earth, he is going to assign mankind as a caretaker of Earth. He created man in the best posture, made him to walk upright while other creatures walk with four legs. He carries man in both lands and seas by means of horses, mules, boats, and ships. He has provided with the goodness of all things like food by means of agriculture, milk by means of cattle, fine clothes, difference in size, shape, colours so that we can distinguish ourselves based on geographies.

On top of everything, God created human beings with a brilliant ability of innovation. Innovation is the indigenous thing of human being which is lacking in other creations. The ability to innovate new things. One of the luminaries of physics, Einstein told "The true sign of intelligence is not knowledge but imagination."

Human being is the only intelligent species on Earth because other animal species don't imagine in the way how we imagine. It is our imagination which paved the way to different languages for communication, different cultures, various fields of maths, science, arts and technology. Curiosity is the sole reason to steer us all the way to step foot on the surface of the Moon.

It is inbuilt in our DNA to be intelligent. The scent of roses you smell, the colours you see, the book you are holding, the sounds of birds you hear everything happens inside a pitch-dark room of skull with the help of a single meat piece called brain. Human brain is a remarkable part in our body. Have you ever thought of how much work your brain does without you even realising it? When we burrow deep inside our brain, the more the treasures it reveals. Human brain is a God's great magnum opus.

Brain is the most complicated device; we have ever found in the universe. Still there are many mysteries yet to unravel inside the human brain. See, this morning I have taught my son about colouring the picture. While doing it I have my eyes focused on colours must fall on the correct sections, my hands are busy holding the coffee mug, I can feel the warmth of the coffee, I can taste the sweet of the coffee, I can hear the voices of my wife residing outside the room, I can smell the scent of the new colouring book, this scent took myself to my childhood days and bring back my past memories.

Just imagine my brain is doing all these things in paralleled manner without any delay. Our brain is the control centre of our body. There are two states of our brain activities. The first activity which is done in a conscious manner like reading, colouring, thinking, smelling, and learning new things etc... The other state of brain activities which have been done in an unconscious manner like our heartbeat, respiration, and digestion. Our brain involves in all these unconscious activities automatically.

Our brain is in auto pilot mode to do all the unconscious activities without our intervention. If you have look at it in this way more than 90% of our everyday activities happens in an unconscious manner. Our brain is taking care of everything for us. Even you can feel the touch of the feather in your fingertips in a precise manner. Everything we see, smell, hear, touch, think. Our brain receives and sends all the nerve impulses by means of electrical impulses.

Just imagine the task of the brain as a million people asking you different questions from different fields at a time, you have to give correct answers everyone at the right time you must be an expertise in each field in order to give the correct answers. Any one of the incorrect answers will lead to serious illness or even death. This is a sign that our brain undertakes each task with complete intelligence.

If you consider our brain as a house, the brain stem is called the basement. It links the brain with the spine. It takes the responsibility of activities like blood flow, heart rate, respiration attention etc... All these involuntary activities are carried out by the help of brainstem. Think a moment, without the help of brain stem, it is impossible for a human to take the conscious effort to maintain these activities like respiration, heartbeat. We couldn't not give attention to any of our other activities by holding this responsibility. We cannot have a peaceful sleep without the help of this brainstem.

Cerebellum is another part of the brain located at the back and bottom of the brain. It is responsible for coordinating our physical balance and movements of our body. It times all the muscle actions of our body so that our body can move smoothly. You can perform any actions of jumping, walking, running without thinking about your body position. For example, if you are set out for jogging, when you have come across a stone. You no need think about the size and dimensions of the stone so that you can either jump or take a round turn with what height or how much degrees in order to overcome the stone.

Likewise you are going to catch a ball, without your consciousness you can move your arms in a certain degree at the right time, right direction based on the speed and direction of the ball. You no need to waste your time thinking about this. Cerebellum will take care of all the precise calculations of your positioning and balancing of muscles in a smooth manner without your intervention.

Hypothalamus is the other region of the brain. It is located at the base of the brain. The size of the hypothalamus is no larger than a pea. It is buried deep inside the brain. We cannot see it without dissecting the brain. It maintains the body temperature, sensation of thirst, sleeping, appetite, working of all hormones in the body. It secretes various hormones which reaches every part of our body and carry out its vital functions. Think for a moment, when you were a baby, your weight was around three kilograms, after becoming an adult your weight has been around 80 kilograms with the same parts of your body. What mechanism makes you to attain this weight?

The reason behind this is growth hormone. Growth hormone makes some cells divide and multiply. For some other cells, it simply increases its volume without multiplying it. For example, The size of a baby heart is 1/6 times the size of an adult heart. But the number of cells present in the heart of a newborn baby and the heart of an adult is same 60 million cells. The growth hormone commands each cell to extend its size during the developmental stages rather than divide and multiply. Thus each cell inside the heart grows and eventually reaches the adult size. Growth hormone concerning the growth of all the part of the body. Hypothalamus constantly secretes such hormones and spreading throughout the entire body.

Despite these hormones travels throughout the entire body, it affects only the relevant parts of the body. It reaches the necessary parts at the right time, right speed and made changes only at the required level. For example, it causes our hair to grow at the right level neither more nor less. These changes varies with gender. For men, these hormones have to cause the beard to grow, deepen the voice. For female, these changes should be prevented. Hormones work in our body in a conscious and intelligent ways. Hypothalamus of our brain regulates the secretion of our body hormones.

Cerebrum, which is the largest part of the brain. It makes up 85% weight of our brain. It is located at the uppermost part of the brain. It is divided into two hemispheres: Right and left hemispheres. Each controls the opposite side of the body. For example, If a stroke occurs on the right side of the brain, the left arm or the left leg of the particular person become weak and paralysed. The function of the cerebrum is interdependent. Each part of cerebrum has its own function. Some perceives light and gives you in the form of image, some convert the outside sound into speech, some register the smell, some helps you to remember a familiar face etc... since there is a division between the two halves of the brain. It is very important to transfer the information from one side of the brain to other side.

It is ensured by means of a bridge called corpus callosum. It is a bundle, made up of millions of nerve cells. This information transfer must occur at lightning speed between the two divisions, if there is any fraction of second delay it would cause a great chaos in our senses. For example, if you see an image by one eye, it would not be compatible with the other eye. It gives kind of a double vision. We cannot hear sound at the same time by using our both ears. We will feel echo if there is delay in the information transfer between two sides.

The backbone or the spinal column provide the major support to our body. It is allowing us to stand upright, bend and twist. The spinal nerves transmit the message between spinal cord and the rest of the body including the muscles, skin, and the internal organs of the body. It is the main pathway of the body's communication to the brain and the brain sends command to the other regions of the body. Sometimes, spinal column works in an independent way, without taking control from brain. Our body has an automatic safeguard mechanism to help us in emergency situations. It is called our reflex actions.

Although, brain is our command centre, there is an emergency pathway in our nerves which command our body to act very quickly before the signal reaches the brain. Most of our reflex actions are directed by the nerve cells in our spinal cord. It gives us fast reflexes in a short path. If you step foot on a sharp object, the cells responsible for your sense of touch immediately sends signal to the spinal cord. The signal coming back from the spinal cord caused us to lift the feet immediately without any time delay. So that our organs can be saved without much injury. This signal reaches the brain and analyses later. Even the slightest alterations in the speed cause serious consequences.

Our brain has the most complex network in our body. Nowadays we can use internet to send or receive messages all over the world with just a click away. Our brain network is far more complicated and accurate than today's internet connection. Inside our brain, there are number of nerve cells connected to one another with many links. There are 100 billion neurons in the brain. It has trillion of connection points in the brain, which serves the efficient connection between the body and our brain. If you start counting the nerve connection of our outer surface of the brain alone at the rate of one connection per second, it will take you million years to complete.

But still there is no traffic and no confusion among the communication between the nerve cells. The number of neuron connections inside the brain is way beyond the number of stars present in our Milky Way galaxies. But the area covered by our brain to maintain this connection is very small. All connections between the nerves cells transmit data all over the body with the speed of 100 metres per second without any delay or confusion. Each neuron has a specific length. For example, the neuron transmits message to your toe is little, shorter, you will not feel any sensation, then you would not retract your feet while stepping on sharp objects. It would cause more harm to our organs. There is not even one of the trillion connections goes in wrong direction. There is a major difference between the nerve cells and the other cells of our body.

The ordinary cells can divide, die, and get replaced. But the nerve cells cannot be renewed. If it gets damage, it cannot be replaced. Throughout our life span, we lose millions of neurons, it will not get replaced. But our every action leaves a mark in the connections, our brain works based on that. So, every one of our brains is unique like a fingerprint. You can feel the cool breeze flows on your face, feel the pressure of the book you are holding, you can see every object in a three-dimensional view, you can feel hunger, thirst everything depends on the flawless transmission between the connections. It is impossible for a human to establish such a complicated connection in a flawless manner.

For all the sensation you feel, even the cool air touches your face, by means of these connections it will be transmitted into electrical signals to your brain. The brain monitor and controls your body in terms of electrical signals. Everything you see, smell, touch, taste it gets transmitted into electrical signals and processed in brain. Despite the brain occupies very small space in our body, while working it consumes sugar and oxygen ten times faster than the other cells in our body which are at rest.

Even if the brain is deprived of oxygen for five minutes, it would leads to brain damage. Just imagine how much electrical energy our brain can spend for all these necessary activities; it is very economical than our household electrical appliances. Our brain has the sufficient electricity to light a 20-watt bulb. It is doing enormous activities, every second with the help of electrical signals in a very economical way. It is very efficient; it doesn't waste energy while carrying out these activities.

We are in the age of advanced technology, which imitates the human brain to the computer. But our brains are 100,000 more times efficient than computers. Our DNA can hold more data than any of our digital devices. Although the invention of computer makes our life easier in all ways, our brain has far more superior systems than the computer.

Because of the complex communication network inside our brain, scientists believe that even the most advanced computer of today's age is still a basic version of our human brain. Our brain takes care of our emotions, sleep, breathing etc...

An astonishing fact about our brain than computers is, our brain constantly changing and learning new things. It can reconstruct its own activities. For example, due to accident a person got brain damage, unharmed part of the brain take care of the activities performed by the damaged brain region. This kind of activity does not present in the computers. Our brain will not do every activity in a preprogrammed way like computers.

For example, if you wear clothes or watch, for the first time you nerve cells sends signals to the brain you will feel the weight of the watch, softness of the clothes, pressure etc... but soon our nerve cells stops sending signals to the brain. Our receptors stop reporting the constant stimuli to our brain. So, soon we forgets the weight of the watch or our clothes. Our receptors adapt it to the constant stimuli. If it keeps on sending signals to the brain like the first time, you will feel a kind of discomfort all the time you wear something things. Like many ways, our brain adapt itself based on the situation in a very conscious and wise manner.

From the first man, mankind have witnessed different era, different ages of the world. There was an age of magic. Egypt was the homeland of world's best magicians. Ancient people were amazed by the tricks of magicians. They have turned their wooden stick into moving snakes, magic helps barren women to conceive, they can interpret dreams and predict the future, using magical spells they offer remedies for long term illness. They strongly believe in supernatural forces; they thought this materialistic world could not bring solution to their everyday problems.

They resort to magic for all their situation of life. They cannot imagine a life without magic. It like science for them. In the same way, we are seeing today's technology. We are believing in scientists that soon they will gain the power to create living or animate beings resembling human. But God has created mankind with unique form of intelligence, knowledge and understanding that differentiate us from all the seen and unseen creatures of the universe. To bring Adam to life, God blow his soul into the lifeless body of Adam.

We can make artificially intelligent computer with installing new programmes and devices. It can be capable of hearing, seeing, speaking, demonstrating intelligence. Yet it doesn't make it into a living being and more certainly we cannot create anything with souls. There is a divine purpose behind the creation of mankind. As we have discussed so far, we are utterly dependent on our creator in every moment. If we show gratitude, he will bestow more of his blessings in this world and the hereafter. This is a journey from God to God. Our life is too short, so be concerned about the things what you are bringing before Almighty on the day of judgement.

The deep you know about his creations, the more you understand his might. He made all the worldly things subservient to us. He is one brought the Moon under our feet. Anything is possible, nothing is impossible for our creator. Dream big, stay positive, spread smiles, forgive people, firm your faith, reach greater heights, celebrate your life, embrace magnificence every moment of your life.

References

- ***Anthrax-the threat***, Centre of disease control and prevention
- ***The science of defending sleepwalkers that kill,*** published in BBC news on 25 March 2010
- ***Dr. Masaru Emoto and water consciousness*** posted in the wellnessenterprise.com
- ***The power of supplication-*** true story by Mufti Ismail Menk
- ***How fatigue played a role in some of the world's biggest disaster*** published in www.optaalert.com
- ***Positive Therapeutic effect of Intercessory prayers*** published in National Library of Medicine (NLM)
- ***Ancient Egypt Magic*** by Geraldine pinch